"*Hydro Plant Electrical Systems* is an indispensible, one-of-a-kind publication that provides the most current information in its field...an excellent reference that encompasses the major electrical components, inspection, testing, and commissioning, which would normally take a multitude of standards and many years of hands-on experience to replicate....[E]asily understood, from the novice to the experienced professional."

— *Daniel L. Purzycki, P.E.*
Vice President of Engineering
North American Hydro
Schofield, Wisconsin, USA

"The book will defend its position as a reference document for mutual understanding not only between engineers of different fields of electrical engineering, but also between engineers of different professions—e.g., civil and mechanical works engineers involved in hydropower design."

— *Øivind Nicolaysen*
Senior Adviser, Electrical Systems
Norconsult International AS
Sandvika, Norway

"Through this book, Mr. Clemen has provided a great service to those who own and operate hydro plants. It is a well organized, most comprehensive and complete reference source, one which should be in the library of every individual involved in specifying, procuring, or operating hydro plant electrical apparatus."

— *T. Earl Goodeve, P. Eng.*
Vice President, Operations
Iris Power Engineering
Etobicoke, Ontario, Canada

"I congratulate David Clemen for presenting his great design knowledge in an easy, understandable, and comprehensive way.The book provides a very good description of the design of hydropower stations... an excellent manual for considerations in designing, testing, and commissioning."

— *Willy Lord*
Senior Expert of Power Installations
SwedPower (A Vattenfall Group Company)
Stockholm, Sweden

MORE EXPERT COMMENTS ON BACK COVER

Hydro Plant Electrical Systems

BY DAVID M. CLEMEN

PUBLISHED BY

HCI PUBLICATIONS

Printed in the United States of America
ISBN 0-9651765-3-3

NOTICE

Neither the author, HCI Publications, nor any person acting on behalf of them (a) makes any warranty, express or implied, with respect to the use of any information, apparatus, method, or process disclosed in this book or that such use may not infringe privately owned rights; or (b) assumes any liabilities with respect to the use of, or for damages resulting from the use of, any information, apparatus, method, or process disclosed in this book.

HCI Publications is the publisher of *Hydro Review* and *HRW–Hydro Review Worldwide* magazines, *Hydrowire* and *HydroWorld Alert* newsletters, books, and other materials useful to individuals and organizations having interests in hydroelectricity. HCI also organizes hydro-related meetings and events, including the HydroVision conferences.

HCI Publications
410 Archibald Street
Kansas City, MO 64111

Phone: (1) 816 931 1311
Fax: (1) 816 931 2015
E-mail: hci@aol.com

Foreword

Electrical engineering at hydropower plants is a highly specialized topic yet, at the same time, one to which every hydropower professional is exposed to some degree. At the heart of any hydropower enterprise is the generation and delivery of electricity, the technical aspects of which are fundamental influences on the day-to-day operation, maintenance, civil design, and economics of facilities. In addition to the basic production of electricity, the field of electrical engineering encompasses some of the most rapidly-evolving new hydropower technologies, including computerized control systems.

This book, *Hydro Plant Electrical Systems*, has its origins in a series of articles that I authored for publication in *Hydro Review* and *HRW* magazines from 1989 to 1998. Individually, the articles cover specific topics in hydro plant electrical engineering. Taken together, they cover a wide range of issues encountered in constructing, operating, or rehabilitating hydroelectric projects. Because of this diversity, HCI Publications suggested that I compile and enhance the articles in book format. This effort entailed considerable technical updating, as well as the preparation of new material covering generators, transformers, controls and protective relays, surge protection, battery systems, unit and station auxiliaries, and start-up and commissioning.

Numerous individuals have made essential contributions to this book. I would like to thank HCI Publications' staff. Among HCI staff, Carl Vansant, Editor-in-Chief, has championed the idea of publishing the book and has overseen its development. Managing Editor Marla Barnes and Technical Editor John Gulliver were responsible for shaping much of the material which originally appeared as magazine articles. Contributing Editor Ellen Faulkner has been the principal editor of this book, and her efforts are sincerely appreciated.

I am most grateful to my professional colleagues who reviewed the manuscript; their comments have led to many improvements.These individuals include Willy Lord of SwedPower (Sweden), Bert Milano of the Bureau of Reclamation (U.S.), Øivind Nicolaysen of Norconsult International AS (Norway), Tapio Pärssinen of Fortum Engineering Ltd. (Finland), and Paul Micale of the Tennessee Valley Authority (U.S.) Paul is the chairman of the Hydroelectric Power Subcommittee of the Institute for Electrical and Electronic Engineers.

Also, I wish to thank the following individuals who reviewed the final manuscript: Brad Berreth of Current Design, LLC (U.S.), Earl Goodeve of Iris Power Engineering, Inc. (Canada), and Dan Purzycki of North American Hydro (U.S.).

Thanks are especially due to Harza Engineering Company. It is on Harza assignments that I have gained the majority of the experience that I am able to share in this volume. In enabling me to put forth the extra effort that has led to publishing this book, I am particularly grateful to Ed Carter, Head of Business Development Operations, Vincent Zipparro, Chief Engineer, and Refaat Abdel-Malek, President and CEO.

David M. Clemen, P.E.
Associate and Senior Electrical Engineer
Harza Engineering Company

About the Author

David M. Clemen, P.E., has worked in the field of power plant electrical engineering for 25 years, 17 of them with Harza Engineering. Now an Associate and Senior Electrical Engineer with Harza, Mr. Clemen has been responsible for design, review, and inspection of electrical systems at numerous conventional and pumped-storage hydroelectric plants, as well as at fossil-fuel plants. In addition to many projects in the United States, his responsibilities have included oversight of electrical design, installation, and commissioning at hydroelectric facilities in the Phillippines, Thailand, Iran, El Salvador, India, Nepal, and Ethiopia.

Contents

CHAPTER 1: **Introduction**

CHAPTER 2: **Hydroelectric Generator Basics**

Types of Generators 2-1
Generator Components 2-4
Generator Ratings 2-9
Excitation Systems 2-11
References 2-12

CHAPTER 3: **Testing Generators Before Commissioning**

Identifying the Required Tests 3-1
Electrical Tests 3-3
Keys to Successful Testing 3-13
References 3-13
Bibilography 3-14

CHAPTER 4: **Scheduled Inspection Outages for Hydroelectric Generators**

Why Have Scheduled Inspection Outages? 4-1
Major and Minor Inspection Outages 4-2
Determining the Frequency of Major Inspection Outages 4-7
Working with the Manufacturer 4-10
Planning Inspection Outages: A Common-Sense Approach 4-10
References 4-11

CHAPTER 5: **Transformers at Hydroelectric Plants**

Generator Step-Up Transformers 5-1
Auxiliary Transformers 5-3
Purchasing Transformers for a Hydroelectric Plant 5-5
References 5-6

CHAPTER 6: Factory and Field Testing of Liquid-Immersed Transformers
Writing Test Specifications 6-1
Factory Dielectric Tests 6-2
Other Transformer Tests 6-13
References 6-15
Bibliography 6-15

CHAPTER 7: Inspection and Maintenance Testing of Liquid-Immersed Transformers
Transformer Inspection and Maintenance Tests 7-1
Insulator Maintenance Testing 7-6
Tests and Inspections of Other Components 7-8
References 7-9
Bibliography 7-9

CHAPTER 8: International Test Standards for Transformers
Standards for Transformer Efficiency (Losses) 8-1
Lightning Impulse Test Standards 8-2
Standards for Switching Surge Impulse Tests 8-4
References 8-5
Bibliography 8-5

CHAPTER 9: Dry Transformers at Hydroelectric Plants
Alternative Liquid-Immersed Transformers 9-1
Dry Transformers 9-2
Measuring the Performance of Dry Transformers 9-5
References 9-7

CHAPTER 10: Controls and Protective Relaying
Choosing a Control System 10-1
Control Systems at Small Hydro Plants 10-2
Control Systems at Medium and Large Hydro Plants 10-5
Equipment Architecture and Protocols 10-8
Protective Relaying 10-9
References 10-14
Bibliography 10-14

CHAPTER 11: Testing Unit Control Switchboards in the Factory and Field

Specifying the Tests 11-1
Importance of Factory Testing 11-4
Field Testing 11-5
Typical Problem Areas 11-6
Considerations at Large Hydro Plants 11-9
References 11-10

CHAPTER 12: Surge Protection

Direct Lightning Protection 12-1
Schemes for Lightning Protection 12 5
Determining Ground Resistance 12-10
Protecting Hydro Plant Control Equipment from Surges 12-11
References 12 21
Bibliography 12-22
Standards 12-22

CHAPTER 13: Batteries at Hydropower Plants

Battery Types 13-1
Selecting and Installing Battery Systems 13-4
Battery Chargers 13-7
Avoiding Common Problems 13-10
Battery Testing Procedures 13-11
References 13-12
Bibliography 13-13

CHAPTER 14: Uninterruptible Power Supply Systems

Types of UPS Systems 14-1
Choosing a System Configuration 14-4
Sizing Hydro Plant UPS Systems 14-6
Writing Specifications for a UPS 14-7
References 14-10

CHAPTER 15: Bus Systems

Standards for Bus Systems 15-1
Choosing Bus Systems for a Hydro Plant. 15-2
Testing Bus Equipment 15-4
References 15-5

CHAPTER 16: Low-Voltage Switchgear for Hydropower Plants

Using LV Switchgear at Hydro Plants 16-1
LV Switchgear Operation 16-3
Breaker Ratings 16-5
Protective Functions 16-6
Upgrading LV Switchgear 16-8
Selecting a Supplier. 16-8
References 16-10
Bibliography 16-10

CHAPTER 17: Start-Up and Commissioning of a Hydroelectric Project

Stages in Start-Up and Commissioning 17-1
Ensuring Successful Commissioning 17-4
Engineering Review During the Commissioning of a Small Hydropower Plant. 17-6
References 17-11

SOURCES

INDEX

Abbreviations and Acronyms

A . . . . amperes
AC. . . . . alternating current
ANSI . . . . American National Standards Institute
AVR. . . . . automatic voltage regulator
AWG . . . . American wire gage
BIL . . . . basic (impulse) insulation level
BMS . . . . battery monitoring system
CMMS. . . . . computerized maintenance management system
CT. . . . . current transformer
CW . . . . chopped wave (impulse test)
DC . . . . direct current
DCS. . . . . distributed (computer) control system
DF. . . . . dissipation factor
EL-CID . . . . electromagnetic core imperfection detector
EO. . . . . electrically operated (circuit breakers)
EPRI . . . . Electric Power Research Institute
FOW . . . . front-of-wave (impulse test)
FOW . . . . forced oil/water (transformer type)
FW. . . . . full wave (impulse test)
GSU . . . . generator step-up (transformer)
HVAC . . . . heating, ventilation, and air conditioning
HPOL . . . . high-pressure oil lift
HTH . . . . high temperature hydrocarbons
IDEC. . . . . Interutility Data Exchange Consortium
IEC . . . . International Electrotechnical Commission
IEEE . . . . Institute of Electrical and Electronics Engineers
ISO . . . . International Organization for Standardization
I/O . . . . input/output
kHz . . . . kilohertz
kV . . . . kilovolt
kVA . . . . kilovolt-ampere
LV . . . . low voltage
MCCB. . . . . molded case circuit breaker
MHz . . . . megahertz

MMI . . . . man-machine interface (software)
MO . . . . manually operated (circuit breakers)
MPT . . . . main power transformer
MTBF . . . . mean time between failures
MVA . . . . megavolt-ampere
NDE . . . . non-destructive examination
NEC . . . . National Electric Code
NEMA. . . . . National Electrical Manufacturers Association
NFPA . . . . National Fire Protection Association
OSI . . . . open systems interconnections
PC . . . . personal computer
PCS . . . . plant control switchboard
PF . . . . power factor
PLC. . . . . programmable logic controller
psi . . . . pounds per square inch
PT . . . . potential transformer
PTFE. . . . . polytetrafluoroethylene
PWM. . . . . pulse width modulation
RCW . . . . reduced chopped wave (impulse test)
RFW . . . . reduced full wave (impulse test)
RMS . . . . root mean square
rpm . . . . rotations per minute
RTD . . . . resistance temperature detector
SAT . . . . station auxiliary transformer
SCADA . . . . supervisory control and data acquisition (system)
TIF . . . . telephone influence factor
UAT . . . . unit auxiliary transformer
UCA . . . . utility communications architecture
UCS . . . . unit control switchboard
UPS. . . . . uninterruptible power supply
VAR. . . . . volt-amperes reactive
VDC . . . . volts direct current
VFC. . . . . variable frequency control
VPE. . . . . vacuum pressure encapsulated (transformer)
VPI . . . . vacuum pressure impregnated (transformer)
WDF . . . . waveform deviation factor
WSCC . . . . Western States Coordinating Council

Chapter 1
Introduction

This book, *Hydro Plant Electrical Systems,* provides an overview of existing practice in electrical engineering at hydroelectric projects. Rather than focussing on design of specific electrical components, the emphasis is on practical aspects of selecting, installing, testing, and maintaining the equipment. The material covered represents both the practical experience of the author and procedures documented in various industry standards, journal articles, and other sources.

The book focuses on standard practices, as well as some newer practices which appear to be growing in popularity in hydropower applications. However, there are very few "textbook" hydroelectric plants. Every one has its own constraints and challenges. These may arise from the physical setting of the project, the historical design and previous installation of equipment, local regulations and engineering traditions, environmental issues, or economic limitations. No engineering project, electrical or otherwise, should be undertaken without an understanding of these factors. Often, when site conditions force departures from standard practice, the result is a valuable innovation.

Several sections of the book are devoted to testing of electrical equipment in hydro plants, both before commissioning and to assess performance after several years of operation. Numerous standards exist for the testing and evaluation of electrical equipment. This book is in no way an attempt to replace or condense those standards, but is intended to provide guidance on their practical application. The reader will gain knowledge of the testing method and equipment, sources of error and reasonable error tolerances, the parameters being measured, and how to interpret the results. Other chapters address control systems, protective functions, emergency power supplies, and surge protection. The final chapter covers start-up and commissioning of hydroelectric projects—the moment that brings together all of the issues addressed separately in preceding chapters.

The book will be most useful for readers with some familiarity with the general principles and terminology of electrical engineering. However, other readers will find much of the material quite accessible and helpful in understanding the considerations that must go into designing and maintaining hydro plant equipment. In addition to the Reference section, which lists cited works (including applicable standards), many of the chapters contain a Bibliography section that presents additional sources of information for interested readers.

Chapter 2
Hydroelectric Generator Basics

Hydroelectric generators have been installed at powerhouses, papermills, mines, and other facilities from the turn of the century to the present day. Their power output ranges from a few kilowatts to several hundred megawatts.

Hydroelectric generators are salient pole machines. In contrast to the cylindrical rotor machines commonly used at fossil-fuel plants, which operate at speeds up to 3,600 rpm, salient pole generators have relatively slow operating speeds, in the range of 80-1000 rpm. Hydroelectric generators typically are designed with ten to 120 rotor field poles projecting from the rotor shaft. Their operating parameters are defined by the American National Standards Institute (ANSI) and the Institute of Electrical and Electronics Engineers (IEEE) in ANSI Standards C50.12 and C50.13.[1,2]

The design, installation, and maintenance needs of hydroelectric generators vary from one machine to the next, depending on factors such as required output, rotating speed, and manufacturer. Generator design has also changed substantially over the years. For a good general discussion of hydro generator maintenance, the reader is referred to the IEEE's Publication No. 492.[3] Although the IEEE withdrew this standard in 1996 and plans to replace it with an expanded standard in the near future, Publication 492 still contains a wealth of "good engineering practices" related to hydro generator maintenance.

Types of Generators

There are three types of hydro generators, distinguished by design speed and the necessary bearing arrangement.

Umbrella Type Generators

This design, shown in Figure 2-1, is used for low-speed (80-200 rpm) generators. It has a combined bottom thrust and guide bearing, and, due to its low operating speed, does not require a top guide bearing. The elimination of the top guide bearing makes this the least expensive generator design.

Semi-Umbrella Type Generators

This design, shown in Figure 2-2, is used for medium-speed (130-250 rpm) generators. It has a combined bottom thrust and guide bearing and a top guide bearing. The top guide bearing is supported by a light bracket. This bracket and bearing arrangement makes the semi-umbrella design more expensive than the umbrella design.

Standard Type Generators

This design, shown in Figure 2-3, is used for high-speed (170-1000 rpm) generators. It was the standard generator design prior to the introduction of the umbrella and semi-umbrella generator designs in the late 1930s. This design has a top-mounted thrust and guide bearing supported on a heavy bracket, which must be capable of supporting the total weight of the generator without distortion during unit operation. This is the most expensive generator design, and consequently is only used for high-speed generators, which have small rotor diameters and inadequate space below the generator to accommodate a combined thrust and guide bearing.

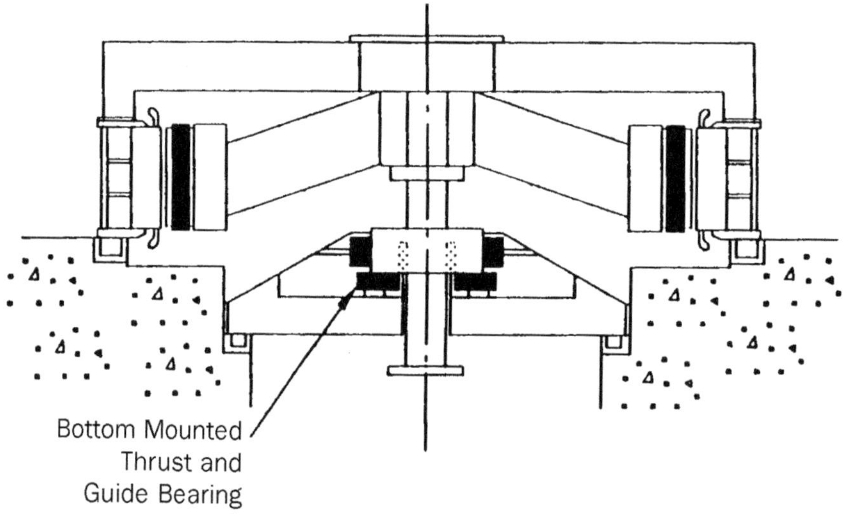

FIGURE 2-1: ***The Umbrella Type Generator.*** *The umbrella type generator, designed for low operating speeds, does not require a top guide bearing.*

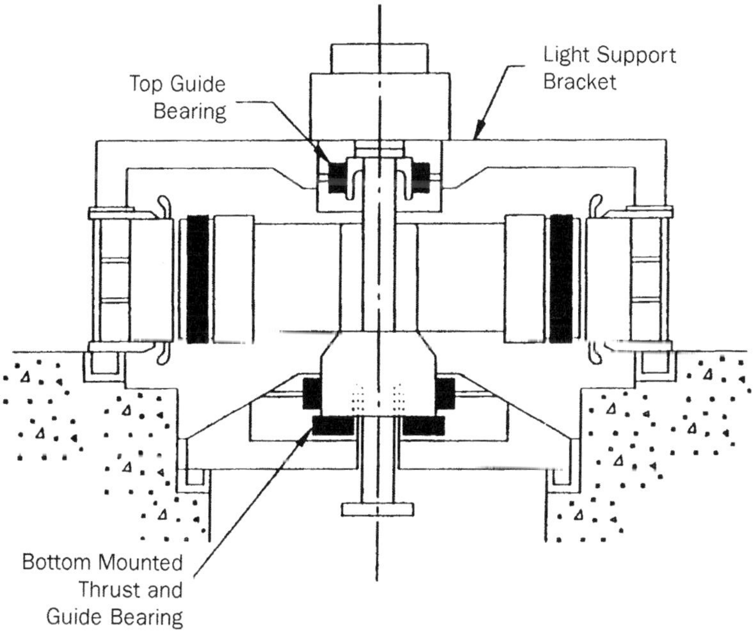

FIGURE 2-2: *The Semi-Umbrella Type Generator.* *The semi-umbrella type generator has a combined bottom thrust and guide bearing and a top guide bearing.*

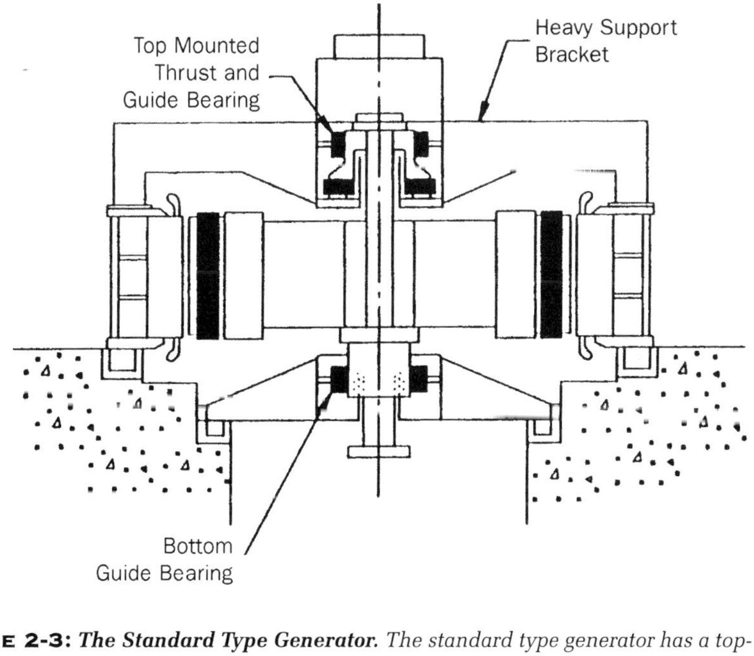

FIGURE 2-3: *The Standard Type Generator.* *The standard type generator has a top-mounted thrust and guide bearing, which supports the entire weight of the machine during operation.*

GENERATOR COMPONENTS

A hydroelectric generator consists of several major components. Figure 2-4 shows the various component parts for a standard type (high-speed) generator. Each major component part is described in the following paragraphs.

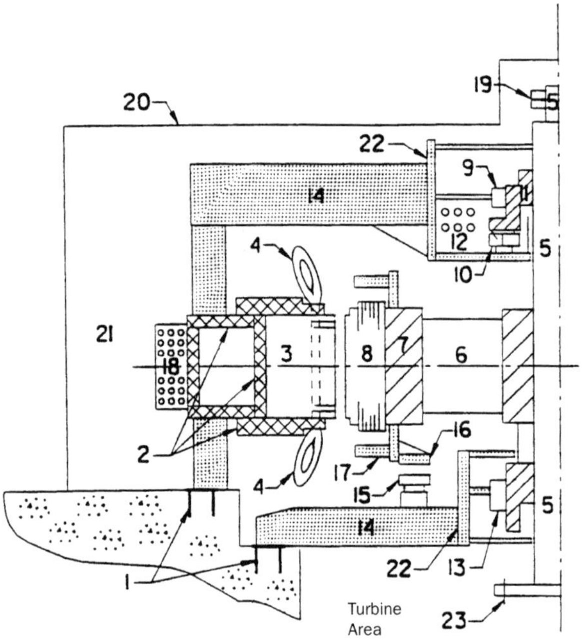

1. Stator Foundation Support Members embedded in concrete
2. Stator Frame
3. Stator Core
4. Stator Winding
5. Shaft
6. Rotor/Spider
7. Rotor Rim
8. Rotor Poles
9. Upper Guide Bearing
10. Thrust Bearing
11. Thrust Bearing Collar
12. Thrust Bearing Cooler
13. Lower Guide Bearing
14. Bearing Brackets (upper and lower)
15. Brake Pad
16. Brake Ring
17. Ventilation Fans
18. Air Heat Exchangers
19. Slip Rings
20. Housing
21. Maintenance Walkway Area
22. Bearing Housing
23. Bolted Connections to Turbine Shaft

FIGURE 2-4: ***Cross Section of a Standard Type Hydroelectric Generator.*** *This figure shows the typical arrangement of components for a standard type hydroelectric generator. The umbrella and semi-umbrella types of hydroelectric generators are similar to the standard type, except for the bearing and bracket arrangements.*

Stator

The generator stator consists of stator foundation support members, a stator frame, stator core, and the stator windings. The foundation members are embedded in the powerhouse concrete and provide support for the stator frame structural members. The stator frame structural members provide support for the stator core and other machine components.

The stator core consists of thin sheet steel laminations stacked on top of one another. The stator core is secured to the stator frame structural members to avoid mechanical vibrations during generator operation. The cores on larger generators are sometimes secured by radial dowels that connect the stator core to the embedded stator soleplates. The stator soleplates are an integral part of the foundation support members.

The stator windings are inserted into the stator core slots and terminated on collector rings, commonly called the ring bus, located at the top of the stator core. The collector rings are routed around the outside periphery of the generator, and are used to connect parallel circuits of a given phase together and to act as the neutral element of the winding. The windings may consist of multi-turn coils or single turn bar windings depending upon the size of the generator. Traditionally, multi-turn coils were used in hydro generators. Recently, however, there has been a trend toward lower-cost single-turn wave windings with stator bars when conditions allow.

Rotor

The salient pole type rotor and its components are shown in Figure 2-5. The rotor hub is attached directly to the generator shaft. The purpose of the rotor hub is to mount and fasten the rotor spider. The rotor spider consists of several mechanical arms that radiate outward from the rotor hub. A continuous rotor rim is connected to the outside end of each rotor spider arm. In physically large generators, this rim consists of individual steel sheets, stacked and bolted together to provide a continuous outside rotor perimeter to which the rotor poles are fastened. In smaller generators, the rotor rim is made of a solid forged steel ring. In small, low-speed machines, the rotor poles are bolted directly to the rim. In larger machines with high or medium speeds, the rotor poles are connected using a dovetail or T-head connection.

The field windings are wrapped around the outside of each rotor pole and connected to each other to form alternate north/south poles around the periphery of the rotor. Historically, due to the high peripheral velocities of the rotor poles, a continuous field winding commonly called "strip-on-edge" was the only acceptable type of field winding. This was a continuous flat copper strip wound edgewise around the pole body. The continuity of the strip prevented lateral stress fractures or creepage of the field winding. However, the "strip-on-edge" fabrication was a very labor-intensive and expensive process. Modern machines usually use flat, rectangular copper strips with brazed dovetail or jigsaw joints at each corner. This process has proven to be reliable in operation and inexpensive to manufacture.

The damper windings consist of copper bars that extend the length of the rotor pole face. The damper windings are used to "dampen" or restrict generator oscillations during steady-state operation. The damper copper bars on each rotor pole are connected to each other. On most hydroelectric generators, the damper windings also are connected from pole to pole.

Shaft

The generator shaft is bolted directly to the turbine shaft and conveys the mechanical power from the turbine to the generator rotor. The generator is

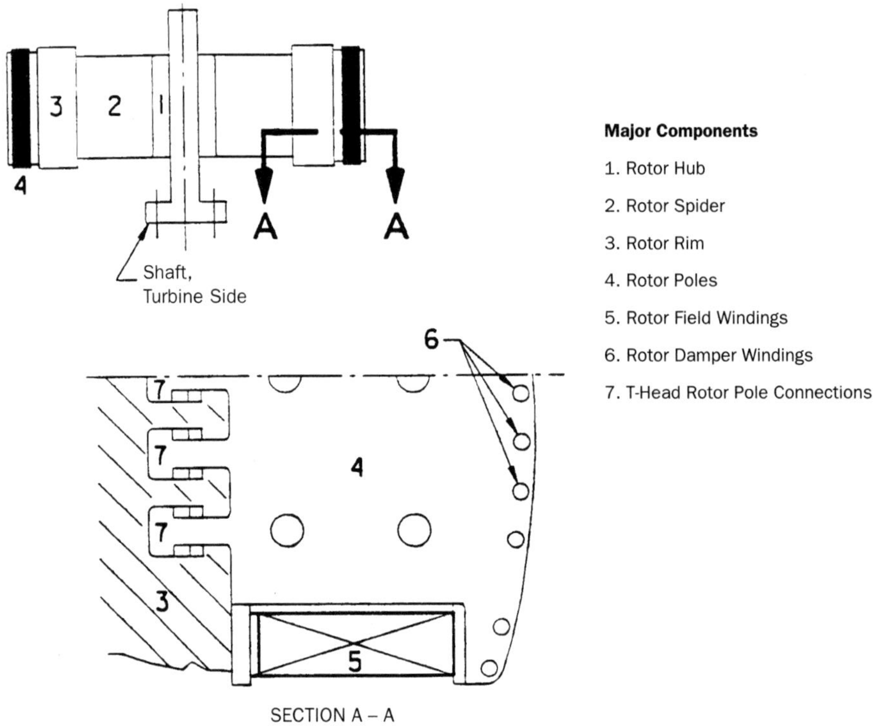

FIGURE 2-5: ***Hydroelectric Generator Rotor.*** *The rotor shown here is for a semi-umbrella type generator. Its components are similar in structure and function to those found in standard and umbrella type generators.*

---

designed with the shaft as short as possible, in order to to reduce vibrations and minimize cost.

Brakes

Hydro generators are provided with a mechanical friction braking system. This system helps stop the generator's rotation after the unit is tripped off-line. Mechanical friction brakes usually are applied when the unit rotation has slowed to less than 25 percent of the operating speed. Stopping the unit at this point avoids wear on the thrust bearing.

Mechanical friction brakes also are applied when the unit is at rest to maintain the unit at standstill. The additional restraining force of the mechanical brakes is required for Francis and Kaplan type turbines if a significant amount of leakage water passes through the closed wicket gates. Water leakage may become a problem after a number of years of unit operation because the seals on the wicket gates become worn. Without brakes, leakage water continuously flowing through the wicket gates causes the turbine and generator to rotate at a slow speed (commonly called unit creep), damaging the thrust bearing.

Modern hydroelectric generators use electrical dynamic braking in addition to the mechanical friction braking system. The electrical dynamic braking system minimizes the wear on the mechanical brake ring and brake pads, prolonging their life. In addition, electrical dynamic braking reduces the duration over which the mechanical brakes are applied during a unit stop sequence, thereby minimizing the amount of brake dust produced by the mechanical brake system. The extended life and reduced brake dust are especially significant for units that are started and stopped several times a day; or units, such as those connected to Pelton turbines, that operate at very high rpms.

Electrical dynamic braking is initiated at a higher unit rotational speed, normally 50 percent of rated speed, and maintained until the mechanical friction brakes are applied. When dynamic braking is utilized in conjunction with the mechanical brakes, the mechanical brakes normally are applied at ten to 15 percent of rated speed. Modern hydroelectric generators, especially pumped storage units, also utilize a brake dust vacuum system to capture most of the brake dust produced when the mechanical friction brakes are applied.

Thrust Bearing

Hydro generators have thrust bearings located either at the top (standard type) or bottom (semi-umbrella or umbrella type) of the generator to support the rotating weight of the machine. There are four major types of thrust bearing design:

- Tilting pad bearings;
- Leveling plate self-equalizing bearings;
- Spring bed mattress type bearings; and
- Self-compensated diaphragm bearings.

The tilting pad type bearing (also commonly called the swivel type) supports each bearing pad on a pivot or adjustable jack screw that is free to tilt. The pivot or jack screw, which allows the bearing pad to tilt freely during operation, is adjusted at the site to ensure that all bearing pads or shoes are set in the same plane. The leveling plate self-equalizing type of bearing supports the bearing pad on a system of interacting levers called leveling plates. The leveling plates interact to equalize loading automatically. Tilting pad and leveling plate self-equalizing bearings are commonly called Kingsbury bearings in the U.S. and Michell bearings in Europe, after the original manufacturers of these bearing types.

The spring bed mattress type of bearing consists of a number of helical springs uniformly distributed under each bearing pad. The individual springs are free to tilt and compress as necessary, depending on the generator/turbine axial thrust.

Self-compensated diaphragm type bearings use a bellows or diaphragm filled with oil to support each bearing pad. An oil line interconnects the bellows and allows them to stabilize hydraulically and tilt as required by the unit's axial thrust.

Modern thrust bearings have a high-pressure oil lift system (HPOL), which injects oil into each shoe pad of the bearing during starting and stopping of the unit. This helps establish a hydrodynamic oil film between the rotating (thrust collar) and stationary (bearing shoes or pads) members of the thrust bearing.

The presence of the hydrodynamic oil film minimizes bearing wear. Consequently, during a unit start the HPOL is turned on before the unit starts to rotate. It is usually shut off after the unit speed exceeds 75 to 90 percent of rated speed because, at the higher rotational speeds, the hydrodynamic oil film is self-sustaining. On unit shutdown, the HPOL is turned on when the unit speed decreases to the 75 to 90 percent range and maintained until the unit is at standstill. The maximum wear on the bearing pads occurs at slow speeds, when, due to hydrodynamic effects, the oil film no longer is maintained over the entire bearing surface.

An alternative thrust bearing design that does not require an HPOL system uses bearing pads faced with polytetrafluoroethylene (PTFE). Although PTFE bearing designs are commonly used in Russia and China and have seen some use in other European countries, they are rarely used in the U.S. or Canada.[6]

Coolers

Water-to-air coolers normally are provided for all modern hydro generators rated greater than 5 MVA. The coolers are situated around the outside periphery of the stator core. Generators equipped with water-to-air coolers can be designed with smaller physical dimensions, reducing the cost of the generator. Automatic regulation of the cooling water flow in direct relation to the generator loading results in more uniform machine operating temperatures, increasing the insulation life of the stator windings.

Normally, generators should be furnished with one more cooler than the number required for operation at rated MVA. This allows one cooler to be removed for maintenance without affecting the unit output.

The generator cooling water normally is supplied from the penstock via a pressure reducing station or pumped from the tailrace. In either case, automatic self-cleaning filters must be provided in the cooling water supply lines to avoid frequent fouling or plugging of the water-to-air coolers.

Fire Extinguishing System

The application of fire extinguishing systems for hydro generators is governed by the National Fire Protection Association's (NFPA's) Standard No. 851, "Recommended Practice for Fire Protection for Hydroelectric Generating Plants."[4] NFPA's standards are based on cost-effectiveness of the fire protection system. Cost-effectiveness is determined by comparing the initial installation cost of the fire protection system to the potential cost of lost generation and repair of equipment damage. As the generator MVA size increases, the initial cost of the fire protection system becomes minimal in relation to the lost generation and repair cost. Therefore, all hydroelectric generators greater than 25 MVA are furnished with either a water deluge or carbon dioxide (CO_2) fire extinguishing system, to minimize the damage caused by a fire inside the machine. In the United States, generators 25 MVA or below are evaluated individually to ensure installation of a cost-effective system.

Since the late 1970s, generators have been manufactured with "self-extinguishing" insulation. With this type of insulation, a fire is extinguished once the short circuit (or other heat source) that initiated the fire is removed. However, in many instances short-circuit current will continue to occur inside the generator until it stops rotating. Earlier generator insulation systems were not self-extinguishing. A fire protection system for generators less than 25 MVA usually is not justified on a cost basis if the generator manufacturer certifies that the insulation system is self-extinguishing. However, oil reservoirs for bearing lubrication should also be investigated thoroughly to ensure they are not capable of sustaining a fire.

In the absence of any international standards concerning generator fire protection systems, a wide variety of systems have been developed throughout the world. The commonly preferred fire extinguishing agent in the United States is CO_2. Other countries, including Canada, Japan, and Russia, commonly utilize water deluge systems. Both types of systems have advantages and disadvantages, and the preference is usually based upon tradition, government regulations, and insurance requirements.[5]

Generator Ratings

Generator ratings are based upon the amount of electrical power that must be transmitted from the turbine to the electrical power grid. For a given discharge passing through a turbine, the turbine power output will vary depending upon the reservoir head. When reservoir levels are high, water is driven through the turbines at a greater pressure, producing a greater kW output, than when the reservoir is low. The generator may be rated to match the turbine output at maximum head or at rated head. The rated head for the turbine is the normal operating head of the reservoir, which is the condition that occurs 70 to 75 percent of the time during a year. During the remaining 25 to 30 percent of the year, the operating head may vary higher or lower than the rated head, depending on seasonal conditions, and non-power constraints such as flood control, navigational demands, and recreational water requirements.

In general, the generator is sized to accommodate the turbine output at rated head. Sizing the generator to accommodate turbine output at the maximum operating head of the reservoir will maximize the amount of energy that can be produced during the year because the generator will be able to accommodate the larger turbine kW output when the reservoir level is high. However, the generator will be oversized for normal operating head and the additional MVA capability will be unused during the 70 to 75 percent of the year that the reservoir is at its normal operating elevation. Consequently, unless the plant is designed for peaking power production, economic analyses will dictate that the generator should be sized to accommodate the turbine output at the normal operating elevation of the reservoir (turbine output at rated head).

Setting MVA Ratings

Once the turbine kW output has been determined, the generator MVA rating must be calculated based upon the system power factor required and the generator efficiency. System power factors range from 0.8 to 0.95, and normal generator efficiencies range from 96.5 to 98.6 percent, depending on the MVA size of the generator. The formula to determine the generator MVA rating is:

$$\text{Generator MVA} = \frac{(\text{Turbine Output Power in MW})(\text{Generator Efficiency})}{(\text{Generator Power Factor})}$$

In addition to defining the MVA rating of the generator, the technical specification should also identify the synchronous condenser and line charging MVAR ratings of the generator. These ratings determine the generator's capability to export or import VARs to support the transmission system without overheating the machine. Due to the large physical size of most hydro generators, the normal synchronous condenser capability of the generator ranges from 55 to 65 percent of the nameplate MVA rating; and the normal line charging capability ranges from 80 to 88 percent of the generator nameplate MVA rating. The exact value for a specific generator depends upon the power factor and short circuit ratios selected for the generator.

The relatively large synchronous condenser and line charging ratings of a hydroelectric generator signify that it is capable of exporting or importing VARs to support the electrical transmission system when reactive power is required. This is very important during system operation of a transmission grid.

Selecting the Armature Voltage

The selected generator armature voltage is based upon the MVA rating of the generator. Generator manufacturers have standardized the voltage ratings to minimize the machine ampacity values. The minimization of ampacity values results in a smaller physical size for the associated bus or cable routed from the generator terminals to the generator switchgear. This reduction in size also results in a lower cost for the bus or cable. Common voltage ratings defined in ANSI C50.12 for the United States are 4.16 kV, 4.8 kV, 6.9 kV, and 13.8kV. ANSI C50.12 also states that voltages above 13.8 kV may be desirable in machines of large capacity.[1]

In addition to the ANSI-defined voltages, the manufacturer may be allowed to determine the generator terminal voltage in order to optimize the generator design.

Excitation Systems

Excitation systems supply and regulate the amount of DC current sent to the generator field winding. The response time of the excitation system to real-world events determines the ability of the generator to maintain its rated terminal voltage and contribute to system stability.

The selected exciter voltage is based upon the amount of excitation current required by the generator. The amount of excitation current is determined by the generator field winding resistance, the generator rpm and MVA size. Higher speed generators require a smaller amount of excitation current to maintain their terminal voltage due to their smaller physical size. Standard excitation system voltages defined in ANSI C50.12 are 62.5, 125, 250, 375, and 500 VDC.[1]

Modern excitation systems consist of two major types, static excitation systems and brushless excitation systems. The modern excitation system replaces older brush-type rotating exciters. The brush-type rotating exciters utilized commutators to change the exciter AC output to the DC required by the generator field winding. The amount of DC current supplied to the field winding was controlled by motor operated field rheostats, which had a very long response time.

Static Excitation Systems

The static excitation system does not have any moving parts, but does require heavy current sliprings and brushgear to transfer the DC current to the generator field winding. The static excitation system is the most commonly used excitation system for hydro generators. Static excitation systems consist of two basic types depending upon the speed of generator field suppression required. The full inverting bridge type uses six thyristors connected in a three-phase full wave bridge arrangement. It allows reversed DC voltage to be applied to the generator field to force faster field suppression, thereby quickly reducing the generator terminal overvoltage during a full load rejection. The semi-inverting type uses three thyristors and three diodes connected in a three-phase full wave bridge. The semi-inverting type drives the positive DC voltage to zero during a full load rejection, but does not allow negative field forcing.

Brushless Excitation Systems

Brushless excitation systems mount a revolving AC exciter, armature, and the thyristors required to convert the AC to DC directly on the generator shaft. This eliminates the need for heavy sliprings and brushgear and simplifies maintenance. However, brushless excitation systems require an additional shaft extension for their physical location and extra physical space at the end of the shaft to enable them to be removed.

Brushless excitation systems may be used on small hydro generators of less than 10 MVA where large DC field ampacities are not required. However, the static excitation system is utilized on the vast majority of hydroelectric generators, even in the small MVA range.

Excitation System Controls

Modern excitation systems utilize microprocessor-based digital controllers as automatic voltage regulators (AVRs) to increase the speed of real-time control functions and to perform diagnostic, monitoring, and data logging tasks. Digital AVRs are capable of remote communication with the plant control switchboard, and are extremely flexible due to their software-based setpoint parameters.

The microprocessor-based digital AVR replaces older solid state analog controllers. The older controllers had fixed setpoints, minimal diagnostic and data logging capabilities, and no communication capabilities. The older solid state analog AVR also responded more slowly than the modern AVR to real-world events.

References

[1] "Requirements for Salient Pole Synchronous Generators and Generator/Motors for Hydraulic Applications," ANSI Standard C50.12, American National Standards Institute (ANSI)/Institute of Electrical and Electronics Engineers (IEEE),1982.

[2] "Requirements for Cylindrical Rotor Synchronous Generators," ANSI Standard C50.13, ANSI/IEEE, 1977.

[3] "IEEE Guide for Operation and Maintenance of Hydro Generators," Publication 492, IEEE, 1976 (reaffirmed 1986).

[4] "Recommended Practice for Fire Protection for Hydroelectric Generating Plants," Standard No. 851, National Fire Protection Association (NFPA), 1992.

[5] "Fire Extinguishing on Large Salient Pole Machines," *ELECTRA*, No. 103, January 1996.

[6] Moss, Bill, "Using PTFE-Lined Thrust Bearing Pads," *HRW,* Vol. 5, No. 5, November 1997.

Chapter 3
Testing Generators Before Commissioning

The electrical and mechanical tests required to ensure that a hydroeletric generator is ready for unit commissioning vary from site to site, and depend upon the method of generator assembly required at the site. The various tests that should be performed, both in the factory and in the field, are documented in this chapter. The emphasis is on electrical testing, because the generator is an electrical machine. The chapter also includes a simple explanation of why each test is required and what the test proves concerning the generator components. Tables summarize the tests required for various categories of generators. Only testing for the standard hydroelectric generator (a synchronous, salient pole machine) is covered. Excitation systems are considered separate from the generator and are not covered.

Identifying the Required Tests

One reason for general confusion about generator testing lies in the variation in size and type of hydroelectric stations. Depending on its megawatt rating and size, the generator may be shipped completely assembled from the factory for a small hydro plant, or, for a large plant, it may be shipped from the factory in its component parts and require complete assembly in the field.

The amount of field assembly required has a direct bearing on the quantity and types of factory and field tests required. If the generator cannot be assembled in the factory because of its MW rating or size, most of the tests would normally be performed in the field after generator assembly is complete.

Categories of Generator

The four categories of generator assembly, which depend on the MW rating and size of the generator, are:

Category 1: Factory assembled generators that are shipped to the site completely assembled. Category 1 generators are usually less than 6 MW, but the classification also depends on the rpm selected, unit speed/weight, and the method of transportation to the site.

Category 2: Factory assembled generators that are shipped to the field as two integral component parts, rotor and stator; or three component parts, the rotor and two stator halves. These are usually between 6 and 20 MW.

Category 3: Generators that require field assembly of the stator, but can have the rotor shipped to the site as an integral component part. Category 3 generators are usually between 20 and 50 MW.

Category 4: Generators that require complete assembly of the stator and rotor at the site. These are usually greater than 50 MW.

Specifications for Testing

A very complete and strongly phrased factory and field test program must be included in the specifications. The tests should be explicitly listed, with identification of specific test values where necessary. All factory and field tests should be witnessed by the engineer/owner, and a manufacturer's test certificate furnished. For smaller generating units (less than 6 MW), where a factory witness or inspection is not cost-effective, the engineer/owner should thoroughly review the factory test certificates once they are received from the manufacturer.

FIGURE 3-1: *Completely Assembled Generator (Category 1).* *A 2-MVA factory-assembled generator shipped as an integral unit to the Dillon Dam Powerhouse in Colorado. The generator is being lowered through the powerhouse roof hatch.*

Specific test values should be assigned to each test, specified per Institute of Electrical and Electronics Engineers (IEEE)/American National Standards Institute (ANSI) standards, unless the specific values are predetermined by the generator terminal voltage rating. If the test value is predetermined, a specific test value should only be listed when it is different from what is required by the IEEE/ANSI standards.

Test Procedures

A number of different test methods can be used to obtain the specific test value stipulated. For example, IEEE Standard 115, paragraph 3.2, identifies three different test methods for the dielectric testing of generators: the AC high-potential test, the DC high-potential test, and the very low frequency test.[1] If the engineer or client has a specific preferred test method, this should be stipulated in the specification. Otherwise, the manufacturer's normal test procedure will most likely be the method used.

A central theme to bear in mind for factory and field tests is that the test values recorded should be very similar, regardless of the location; that is, the insulation resistance test values for the stator windings at the site should be comparable to the insulation resistance test values recorded on the factory test certificates. If the test values are not comparable, this indicates that some problem occurred during field installation (either documented or not documented), and closer investigation is warranted.

Tables 3-1 through 3-6 identify the factory and field tests (both electrical and mechanical) required for the four categories of generator assembly. The factor that determines whether rotor assembly must be performed in the field (Category 4) rather than in the factory (Category 3) is the rotor weight compared to the maximum shipping weight limitation for transportation to the site. Naturally, if a powerhouse site has good access (e.g., direct access by barge or ship) the manufacturer will factory assemble the rotor to reduce field labor costs.

The purpose of the factory tests is to ensure that the machine was manufactured properly and was shipped from the factory in mint condition. The main purpose of the field tests is to verify the factory test values after installation of the generator at the site. The field tests ensure that the components of the generator were not damaged in any way during shipping and installation.

Generator manufacturers routinely perform numerous factory tests on the various component parts as part of their quality control procedures. These tests include corona discharge tests on random stator coils to ensure no voids are present in the insulation and rotor pole voltage drop tests to verify that none of the field windings are short circuited. Tests and procedures vary with the individual manufacturer. A complete discussion of quality control tests associated with individual component parts is beyond the scope of this book, and reference should be made to the individual manufacturer and to relevant industry standards (see bibliography). In addition, the reader may refer to the many in-depth technical articles and textbooks that have been written concerning the various electrical tests.

Electrical Tests

The electrical tests described are selected sequentially from Tables 3-1 through 3-3 (factory tests) and Tables 3-4 through 3-6 (field tests). The tables also list mechanical tests, although these are not covered in detail here.

Resistance Test of Armature and Field Winding

This test measures the stator coil winding and field winding copper resistance.

TABLE 3-1: ***Factory Tests for Categories 1 and 2 Generator Assembly***

Category 1 generators are shipped completely assembled from the factory and are usually less than 6 MW. Category 2 generators have the rotor and stator assembled as a complete generator in the factory, but the rotor and stator are shipped individually. The stator may be shipped in half segments. These are usually 6 to 20 MW generators. The same factory tests apply to Category 1 and Category 2 generators.

1. Resistance test of armature and field windings.
2. Dielectric test of armature and field windings.
3. Insulation resistance of armature and field windings.
   This shall include the polarization index values for both armature and field windings.
4. Stator core loop test at rated flux for one hour.
5. Phase rotation check.
6. No-load saturation test.
7. Short circuit saturation test.
8. Mechanical balance of rotor.
9. Dynamic balancing of rotor at 125 percent rated speed.
10. Current transformer test.
11. Efficiency test.
12. NDE tests of rotor shaft and shaft coupling bolts.
13. Material test certificates of various component parts.
14. Temperature rise test.
15. Hydrostatic test of each oil/air cooler, if applicable.
16. Power factor tip-up of stator coil insulation (optional)

TABLE 3-2: ***Factory Tests for Category 3 Generator Assembly***

Category 3 generators have the rotor completely assembled in the factory; complete assembly of the stator (core stacking, etc.) is required in the field. The rotor is shipped from the factory as an integral unit. These units are usually between 20 and 50 MW.

1. Resistance test of rotor winding.
2. Dielectric test of rotor winding.
3. Insulation resistance test of rotor winding, including polarization index test.
4. Mechanical balance of rotor.
5. Dynamic balancing of rotor at 125 percent rated speed (if possible in factory, depending on size of rotor).
6. Dielectric tests of individual stator coil windings.
7. NDE tests of rotor shaft.
8. Material test certificates of component parts.
9. Hydrostatic test of each oil/air cooler.
10. Current transformer tests.
11. Shop alignment check of combined turbine and generator shafts.
12. Power factor tip-up of stator coil insulation (optional).

TABLE 3-3: ***Factory Tests for Category 4 Generator Assembly***

Category 4 generators require complete assembly of stator and rotor in the field, and are generally greater than 50 MW.

1. Dielectric tests of individual stator core windings.
2. NDE tests of rotor shaft and shaft coupling.
3. Material test certificates of various component parts.
4. Hydrostatic test of each oil/air cooler.
5. Current transformer tests.
6. Shop alignment check of combined turbine and generator shafts.
7. Power factor tip-up of stator coil insulation (optional).

The test normally uses a low-voltage DC source and the voltmeter-ammeter method to determine the resistances. The resistance values in conjunction with the no-load saturation and short circuit saturation tests are necessary to obtain the calculated efficiencies of the generator.

Dielectric Test of Armature and Field Winding

The dielectric test can utilize either AC or DC voltages and is commonly called the *high potential test* (or *"hi-pot"*). The AC test stresses the insulation at a value greater than two times rated voltage (usually 2E + 1,000) to ensure that the insulation is capable of enduring temporary over-voltages. Direct voltage testing also may be used. A direct voltage equal to 1.7 times the RMS value of the AC test voltage must be applied.[1]

The high potential test of the stator windings is performed phase by phase. The two phases not being tested are jumpered together and connected to ground. Each phase winding is short circuited to obtain the full test voltage at each end of the winding; that is, in normal operation the neutral end of the winding has a very limited potential stress applied to it because it is the neutral point. Therefore, during the high potential test, we want to ensure that we stress each end of the winding, thereby confirming the insulation's overall integrity.[2]

This test is usually preceded by a visual inspection and the insulation resistance test as it is the most severe test of the insulation.

Insulation Resistance Test of Armature and Field Winding

Commonly called the "*megger*" or *megohmmeter* test, the insulation resistance test measures the DC leakage current of the insulation by stressing the insulation with an applied voltage from 100 to 15,000 volts. Bear in mind that no insulation is perfect. This test measures the milli or microamps that flow through the insulation. From this current reading, and knowing the fixed voltage at which the test takes place, one determines the actual megohm resistance of the insulation. A high resistance value implies that insulation leakage current is minimal and that the circuit may be safely energized.

This test is usually performed prior to the dielectric test, because insulation failure during the insulation resistance test will not cause any significant damage due to the low values of voltage applied.

The polarization index (PI) is the ratio of the insulation resistance value obtained after 10 minutes of testing to the value obtained after 1 minute of testing. The insulation resistance should increase with time, and a PI value of 2 or greater is usually sought.

Stator Core Loop Test

Also called the *ring test* or *magnetizing test,* this test is performed on the stator core before insertion of the stator windings to ensure there are no defects in the stator core steel or assembly of this steel, and to verify the manufacturer's calculated values. If defects in the stator core were not detected until after assembly of the generator, correction could be extremely expensive. In the last decade, the electromagnetic core imperfection detector (EL-CID) test, described below, has become an increasingly common alternative to the loop test.

The traditional stator core loop test test consists of wrapping a number of turns of heavy cable toroidally around the stator core and frame. The cable is energized with a single phase line frequency current, which is increased until rated flux is obtained. This flux value is maintained for a period of time, usually 30 minutes to 1 hour, at which time the core is checked for hot spots caused by excessive burrs or defective interlaminar insulation on the stator core steel. At the same time, measurements are taken of the input voltage and power, and the volts per turn induced in the stator core, to obtain the stator core losses. This value is compared with the manufacturer's previously calculated stator core losses as a check on the machine design.

This test also physically "shakes down" and compacts the core steel. This compaction may result in a loss of core pressure. Maintaining the design core pressure is important, as this pressure maintains the mechanical rigidity of the core while allowing the rotor magnetic field to penetrate into the core steel. If the core loses its mechanical rigidity, vibrating core sheets could damage the stator winding insulation and also produce eddy currents within the core, creating local hot spots. Therefore, the core bolts and core pressure should be checked after the stator core loop test and, if necessary, the core bolts should be retorqued to obtain the design core pressure.

The stator core loop test arrangement also requires a large external power source and considerable field time to wrap the large conductor cables around the stator frame and core.

Electromagnetic Core Imperfection Detector

Since 1989, a new method of testing the stator core has been utilized in the industry. This method is called the "electromagnetic core imperfection detector," or EL-CID. The EL-CID test can be performed with the rotor removed or in place. However, the test is most commonly performed with the rotor removed. This method of testing is not uniformly accepted by all North American manufacturers and agencies as a substitute for the stator core loop test. It has, however, been used at a number of hydro plants.

The test requires that the stator core be excited to 4 percent of the rated flux (as opposed to 100 percent when using the stator core loop test). This reduction in flux density usually requires wrapping approximately 20 turns of No. 12 AWG (American wire gage) toroidally around the stator core. The power required to obtain the reduced flux level is usually no more than 20 amps.

The EL-CID test uses a special sensing coil which is routed over the interior of the stator core bore to detect flux anomalies. The flux anomalies indicate interlaminar insulation deterioration and/or potential core hot spots, which must be further investigated.

Phase Rotation Check

The discussion of interfacing electrical vectors versus the physical machine rotation is always very complex and prone to errors in interpretation. The phasing shown in diagrams in a utility's office files often differs from the field implementation of the phasing scheme by utility linemen. This test eliminates any possibility of connecting the generator out of phase with the existing utility system.

The test should be performed across the generator breaker for small units, and with the appropriate potential

FIGURE 3-2: ***Rotor Assembled and Shipped Separately from Stator (Category 2).*** *The rotor of a 13.3 MVA factory assembled generator ready for transportation to the Moose River Powerhouse in New York. This generator was shipped to the field as two integral component parts, rotor and stator.*

FIGURE 3-3: ***Installing the Rotor.*** *The rotor for the Moose River plant in New York is lowered through the powerhouse roof hatch for installation inside the generator stator.*

MECHANICAL TESTS

The mechanical integrity of the generator is essential. However, the emphasis of this chapter is on electrical tests associated with the generator. No mechanical standards are referenced in the text. For those interested, the following is a brief summary of the mechanical tests associated with the generator.

Factory Testing

In the factory, mechanical tests usually consist of two parts: review of material test certificates, which confirms the purity/composition of the materials used for the various component parts, and non-destructive examinations (NDE), which confirm that there are no flaws, anomalies, or discontinuities in the component parts. The extent of both these tests is dependent upon the generator size, and can vary from a simple review of the rotor material test certificates for a small generator, to the full range of liquid penetrant, magnetic particle, or ultrasonic testing for larger generators. On the larger generators, a hydrostatic test of each oil/air cooler should be performed, in addition to a shop alignment check of the combined turbine and generator shafts to help ensure that field problems do not occur.

Field Testing

The mechanical field tests for the generator consist of:

- A rotational check of the combined turbine/generator shafts to ensure proper alignment;
- Hydrostatic tests of each oil/air cooler;
- Confirmation of proper generator bearing installation/clearances;
- Confirmation of proper generator air gap clearance;
- Brake operational tests; and
- Tests to determine the flow rates and temperatures of incoming/outgoing cooling water for the bearings and air coolers.

In addition, a moment of inertia test (WR^2) and a runaway speed test are performed on some units. Although this is not a commonly accepted practice in NorthAmerica, certain countries, such as Norway, have a requirement for a runaway speed test on every generating unit..

transformer circuits for larger units, to confirm that the vector phase rotation of the generator matches the vector phase rotation of the system.

No-Load and Short Circuit Saturation Tests

The no-load saturation test, also called the *open circuit saturation test*, is performed with the armature terminals open circuited and the field winding excited. The machine is driven at rated terminal voltage. This test confirms that the predicted excitation will produce the predicted terminal voltage.

The short circuit saturation test, also called the *synchronous impedance test*, is performed with the armature terminals short circuited and the field winding excited. The machine is driven at rated speed, and a curve of the armature current versus field current is recorded as the field current is changed from zero to a value which will produce rated armature current.

Both tests are used in conjunction with the winding resistance values to calculate generator efficiency and other essential synchronous machine parameters (synchronous impedance, short circuit ratio, subtransient reactance, etc.).[4]

Current Transformer Tests

All current transformers (CTs), associated with the generator for functions such as protective relaying, are given "routine" tests as defined by ANSI Standard C57.13 [5] These tests confirm that the CTs are functioning as designed. The manufacturer should be required to provide ratio correction curves and excitation curves as a standard part of the tests. The CTs are normally used for the generator differential and generator neutral grounding protective relays. Therefore, accuracy is essential.

Efficiency Tests

There are several different methods for verifying generator efficiency per IEEE 115.[1] These methods are:

- Separate drive method;
- Electric-input method;
- Retardation method; and
- Cooler (calorimetric) method.

The separate drive method and the electric-input method are direct measurements of efficiency, or the ratio of power output to power input. These tests are applied in the factory to smaller machines. However, for larger machines the mechanical driving power needed for direct measurements of input and output power is not available in the factory. In that case, efficiency is determined in the field by use of a *segregated losses* calculation utilizing either the retardation method or calorimetric method.

An accurate generator efficiency is very important. The efficiency value will be used again in the turbine index test calculations, which verify the turbine efficiency.

Temperature Rise Test

Also commonly called the *heat run test*, the temperature rise test is performed on the generator to verify that the component parts do not reach a temperature higher than their insulation can withstand. Most generators are insulated with Class B or Class F insulation, which should not exceed a hottest spot temperature of 130 degrees and 155 degrees Celsius respectively.

The temperature rise for the generator components' parts is identified in ANSI Standard C50.12, Table 1.[6] Values of the temperature rise not to be exceeded are listed and depend upon the method of temperature measurement, whether the generator is directly or indirectly cooled, and the maximum temperature of the fluid (either air or water) used for cooling the machine, among other variables.

The methods of loading the machine to determine the temperature rise vary with the size of the generator. Four methods are listed in IEEE 115 with detailed explanations.[1] For small hydro generators, the test consists of running the generator at rated load, frequency, and voltage with a constant field current until the machine reaches a constant temperature. Temperature readings are recorded every half hour or less to verify that the generator has reached this constant temperature; then the final readings are compared to the allowable temperature rise to confirm that these temperatures are within the tabulated limits.

For larger hydro generators, one of the other three methods listed in IEEE 115 must be used, because running the generator at rated load is not always possible or practical. The reader should refer to IEEE 115 for a more complete explanation of these methods.[1]

Waveform Deviation Factor and Telephone Influence Factor Tests

The waveform deviation factor (WDF) and telephone influence factor (TIF) tests verify the purity of the sine wave that the generator produces. The tests require a spectrum analyzer type oscilloscope at the site.

The WDF test compares the shape of the sine wave produced by the generator with that of a perfect sine wave. The deviation factor of the open circuit terminal voltage should be less than 10 percent of the phase-to-phase voltage. Usually, a distorted sine wave must have a deviation factor of greater than 5 percent to be visible to the naked eye. Methods to analytically determine the WDF are listed in IEEE 115.[1]

The TIF test analyzes the harmonics produced by the generator and uses various weighting factors originally developed in 1935 by Edison Electric Institute and Bell Telephone to ensure that 60-Hertz power generators would not produce harmonics that would interfere or "influence" the communications circuits. The weighting factors to be utilized for modern generators are listed in ANSI Standard C50.13; the test method to obtain TIF values for the generator is identified in IEEE 115.[7,1]

Field Tests

Tables 3-4 through 3-6 list the recommended field tests for the four categories of generator assembly.

The efficiency of a factory assembled generator is documented with a test certificate in the factory; therefore, an efficiency test is not required for Category 1 or 2 generators in the field. However, a no-load saturation test is performed in the field and the developed curve checked against the factory no-load saturation curve. This verifies that the generator parameters have not changed in any way.

A reduced value dielectric test should be performed in the field. The reason for the reduction in voltage for these dielectric tests is that the generator has already passed the full value dielectric tests at the factory and we do not want to needlessly stress the insulation.

Chapter 3

TABLE 3-4: ***Field Tests for Categories 1 and 2 Generator Assembly***

Category 1 generators are shipped completely assembled from the factory and are usually less than 6 MW. Category 2 generators have the rotor and stator assembled as a complete generator in the factory, but the rotor and stator are shipped individually. These are usually 6 to 20 MW generators. The same field tests apply to Category 1 and Category 2 generators.

1. Insulation resistance tests of armature and field windings including polarization index values for both.
2. A reduced value dielectric test of armature and field windings.
3. Phase rotation check.
4. No-load saturation curve.
5. Dynamic balancing of rotor at 125 percent rated speed.
6. Temperature rise test.
7. Waveform deviation factor.
8. Telephone influence factor.
9. Hydrostatic test of each oil/air cooler, if applicable.
10. Rotational check of combined turbine and generator shafts.
11. Operation of brakes.
12. Test to determine flow rates and temperatures of incoming and outgoing cooling water for bearings and air coolers.
13. Overspeed test.
14. Short-circuit withstand test (seldom performed).

A DC dielectric test is performed, rather than an AC dielectric test, because DC testing is considerably less destructive if a failure occurs. In addition, the DC dielectric test equipment is light, portable, and readily available in the field, whereas the high-voltage AC test equipment is large, expensive, and often difficult to locate.

Fire Protection System Tests

Generator carbon dioxide (CO_2) fire protection systems are required by the National Fire Protection System Standard 12.[8] This standard requires that a CO_2 fire protection system obtain a 30 percent gas concentration in the generator housing during the generator deceleration period, or for a period of 20 minutes, whichever is longer. The 30 percent CO_2 concentration will effectively extinguish any combustion that occurs inside the housing while the unit is decelerating to a standstill.

TABLE 3-5: ***Field Tests for Category 3 Generator Assembly***

Category 3 generators have the rotor completely assembled in the factory; complete assembly of the stator (core stacking, etc.) is required in the field. The rotor is shipped from the factory as an integral unit. These units are usually between 20 and 50 MW.

1. Resistance test of field armature winding.
2. Dielectric test of armature winding.
3. Phase rotation check.
4. Stator core loop test.
5. Phase rotation check.
6. No-load saturation curve.
7. Short circuit saturation curve.
8. Dynamic balancing of rotor at 125 percent rated speed.
9. Efficiency calculation using values from 1, 6, and 7 above or an efficiency test.
10. Temperature rise test.
11. Waveform deviation factor.
12. Telephone influence factor.
13. Hydrostatic test of each oil/air cooler.
14. Rotational check of combined turbine and generator shafts.
15. Operation of brakes.
16. Test to determine flow rates and temperatures of incoming and outgoing cooling water for bearings and air coolers.
17. Overspeed test.
18. Fire protection system test.
19. Moment of inertia (seldom performed).
20. Runaway speed test (seldom performed).
21. Short-circuit withstand test (seldom performed).

TABLE 3-6: ***Field Tests for Category 4 Generator Assembly***

Category 4 generators require complete assembly of stator and rotor in the field, and are generally greater than 50 MW.

1. All tests listed in Table 3-1, Factory Tests for Categories 1 and 2 Generator Assembly, except nos. 10, 12, and 13.
2. Telephone influence factor.
3. Waveform deviation factor.
4. Rotational check of combined turbine and generator shafts.
5. Operation of brakes.
6. Test to determine flow rates and temperatures of incoming and outgoing cooling water for bearings and air coolers.
7. Overspeed test.
8. Fire protection system test.
9. Moment of inertia test (seldom peformed).
10. Runaway speed test (seldom performed).
11. Short-circuit withstand test (seldom performed).

CO_2 fire protection systems are normally tested during startup and commissioning, before the initial unit operation. The test consists of running the generator at rated speed while it is deenergized and disconnected from the system. Operation of the fire protection system is initiated by an electrical release that simulates the detection of an actual fire by a heat detector located inside the generator housing. Operation of the CO_2 system also can be initiated by protective relays or generator lockout relays. During the CO_2 release, the following parameters are measured over a 30-minute period:

- CO_2 gas concentration inside the generator housing;
- Rotational speed of the unit; and
- Pressure inside the generator housing.

These measurements are made every 20 seconds for the first three minutes, and at one-minute intervals thereafter.

Keys to Successful Testing

The generator specifications should contain very complete and clearly phrased factory and field test sections. The tests required should be explicitly listed, with specific test values identified when they are different from the ANSI/IEEE standards. Specific test procedures, if required or preferred, should be identified. All factory and field tests should be witnessed by the engineer or owner.

For all generators that are assembled or partially assembled in the factory, the field tests should exactly replicate the required factory tests to confirm that no problem has occurred during installation of the generator. For generators that must be completely assembled in the field, the field tests should be as comprehensive as possible to provide greater assurance that the machine will function properly throughout its lifetime.

References

[1] "IEEE Guide: Test Procedures For Synchronous Machines," IEEE Standard 115, Institute of Electrical and Electronics Engineers (IEEE), 1995.

[2] Gill, A.S., *Electrical Equipment Testing and Maintenance,* Reston Publishing Co., Reston, VA, 1982.

[3] Walker, J.H., *Large Synchronous Machines,* University Press, Oxford, England 1981.

[4] Kosow, Irving, *Electric Machinery and Transformers,* Prentice-Hall, Englewood Cliffs, NJ, 1972.

[5] "Requirements for Instrument Transformers," ANSI Standard C57.13, American National Standards Institute (ANSI)/IEEE, 1978.

[6] "Reqirements for Salient Pole Synchronous Generators and Generator/Motors for Hydraulic Applications," ANSI Standard C50.12, ANSI/IEEE,1982.

[7] "Requirements for Cylindrical Rotor Synchronous Generators," ANSI Standard C50.13, ANSI/IEEE, 1977.

[8] "Standards on Carbon Dioxide Extinguishing Systems," National Fire Protection Association Standard 12, 1985.

Bibliography

"Recommended Practice for Testing Insulation Resistance of Rotating Machinery," ANSI/IEEE Standard 43, 1974.

"Recommended Practice for Insulation Testing of Large AC Rotating Machinery with High Voltage at Very Low Frequency," ANSI/IEEE Standard 433, 1973.

"Recommended Practice for Insulation Testing of Large AC Rotating Machinery with High Direct Voltage," ANSI/IEEE Standard 95, 1977.

"General Requirements for Synchronous Machines," ANSI Standard C50.10, ANSI/IEEE, 1977.

"Motors and Generators," ANSI/National Electrical Manufacturers Association (NEMA) Standard MG1, ANSI/NEMA, 1987.

Chapter 4
Scheduled Inspection Outages for Hydroelectric Generators

Every generator —regardless of age, size, history, or purpose —will benefit from regularly scheduled inspection outages to ensure the operational integrity of the generator. The recommendations of the generator's manufacturer, the owner's past experience, and test results for the specific machine should provide guidance in developing an inspection program and protocol. Although the specific schedule and scope of the inspection program vary from one machine to the next, there are several types of examinations and tests that should take place during any generator inspection outage. An understanding of the general purpose and procedures involved in these tests can be broadly applied to various generator types.

The Institute of Electrical and Electronics Engineers (IEEE) has published a good discussion of hydro generator maintenance, including inspection outages and tests.[1] Although the standard in which this discussion appeared was withdrawn in 1996, it still contains a wealth of "good engineering practices" related to hydro generator maintenance and inspection outages. (The IEEE intends to issue a revised and expanded version of its standard covering maintenance and inspection outages by late 1999.)

Why Have Scheduled Inspection Outages?

The data collected from regularly scheduled inspection outages enable the owner to:

- Make minor repairs or adjustments before problems affect unit availability;
- Repair major generator components in a scheduled and timely manner; and
- Implement a unit upgrade in a prepared, orderly sequence that minimizes unit downtime and lost revenues.

(Note: There is no common definition of the word "upgrade" pertaining to generators. Here, "upgrade" is used to mean "...any change of parts or parameters to prolong the life and/or increase the capacity of the generator." In general, prolonging the life of a generator or returning it to near-new condition is "life extension," and increasing the capacity or enhancing performance is "modernization.")[2]

Inspection outages should not be confused with repair or upgrade outages. Although many owners use inspection outages to perform normal maintenance, the main purpose of an inspection outage is to provide information needed for planning repair or upgrade outages.

The scheduling of outages can be "time-based" or "condition-based," as defined by the Electric Power Research Institute.[3] Inspection outages scheduled strictly on a time basis, which are more commonly used than "condition-based" outages, have the advantage of being planned in advance. This allows the owner to mobilize the personnel and resources required for the outage in an orderly manner. The length of time between outages can be based upon the owner's experience with the generator. The frequency of inspections also can be adjusted in light of the inspection results from the previous outage. The disadvantage of time-based inspections is that they may not detect failure mechanisms that develop over a relatively short time.

Condition-based maintenance initiates inspection outages only when there are signs of a problem with the generator. Detecting a developing problem requires the use of predictive maintenance components. Predictive maintenance components now available include air-gap monitors, vibration monitors, partial discharge monitors, stator core temperature monitors, and winding vibration monitors. Such devices can detect numerous problems inside the generator while it is operating. The advantage of condition-based maintenance is that it will maximize the unit's availability by scheduling inspection outages only when they are essential. The disadvantage is that the monitoring equipment used requires a significant investment, typically from $100,000 to more than $500,000 per generator. Moreover, the monitoring equipment presently available cannot detect every possible generator failure mode.

MAJOR AND MINOR INSPECTION OUTAGES

Scheduled inspections of hydroelectric generators usually are divided into two categories, minor and major inspection outages.

Minor Scheduled Inspection Outages

These outages correspond to the "partial inspection" defined in IEEE Publication 492.[1] They consist of several electrical tests, mechanical pressure tests, and a visual inspection of the various major components. Because a minor inspection outage does not include removal of the rotor, the associated outage period is usually only a week or two.

The electrical and mechanical tests typically are performed at the start of the minor outage to identify problems in specific components. If a problem compo-

Looking for Problems: Visual Inspections During Minor Scheduled Inspection Outages

A minor inspection outage should include a visual inspection of a number of generator components and electrical connections. Using mirrors and turning the rotor during the inspection will help the inspectors thoroughly check for the following:

Stator

- □ Cleanliness; powder and sludge deposits
- □ End wedge tightness
- □ Main lead bushings
- □ Stator end windings:
  - -evidence of movement or fretting
  - -ties between coils
  - -blocks between coils
  - -ties from coils to support rings
  - -insulation uniformity
- □ Winding insulation:
  - -cracks or tape separation
  - -partial discharge areas
  - -bulging or softness
- □ Stator core
  - -waviness
  - -bent or broken laminations
  - -connections to frame
  - -evidence of movement or fretting
- □ Circumferential/collector ring bus insulation and mechanical connections to stator frame

Rotor

- □ Cleanliness; powder and sludge deposits
- □ Top and bottom field connections
- □ Rotor pole face for evidence of heating
- □ Damper bar interconnections
- □ Field winding insulation (top and bottom)
- □ Rotor fastenings, including dovetail keys or wedges that fasten the pole to the rotor rim
- □ Fan blades for cracks or stress fractures
- □ Interpole V-block support for bolt tightness and cracks

Miscellaneous

- □ Brake ring
- □ Brake pads and associated limit switches
- □ Resistance temperature detector leads and connections
- □ Main leads:
  - -surge arrestors
  - -insulators

nent is found, additional tests may be performed during the outage in conjunction with the visual inspection. Typical electrical/mechanical tests for a 100-MVA generator are listed in Table 4-1. (See sidebar for visual component inspection items.)

IEEE Publication 492 recommends that a minor scheduled inspection outage be performed annually for an initial unspecified time period after installation of

TABLE 4-1: ***Electrical/mechanical tests typically recommended for a minor scheduled inspection outage for a 100-MVA generator.***

Component	Test	Purpose of Test
Stator	**Electrical tests**	
	Insulation resistance	Confirms insulation integrity to ground
	Polarization index	Confirms insulation integrity to ground over a 10-minute period
	Partial discharge	Confirms insulation integrity by measuring internal discharges (on specially equipped generators)
	Mechanical tests and inspections	
	Check stator core	Confirms core is compressed to manufacturer's recommended values.
	Check end wedges	Confirms coil is adequately braced or supported in slot (reinsert and epoxy glue wedges if necessary)
	Check heat exchangers at 1.5 times rated pressure	Identifies leaks or plugged tubes that impair cooling of generator
	Check space heaters	Confirms that condensation will not occur within the generator housing when unit is shut down
	Vibration analysis	Identifies mechanical imbalances, misalignments, etc.
Rotor	**Electrical tests**	
	Insulation resistance	Confirms insulation integrity to ground
	Polarization index	Confirms insulation integrity to ground over a 10-minute period
	AC pole drop test	Detects shorted turns on rotor field windings
	Mechanical tests	
	Air gap measurements (top and bottom at 4 compass poles)	Confirms that air gap clearance between rotor and stator is uniform and complies with manufacturer recommendations
	Check bolt torques, tightness on field interpole connections	Confirms mechanical integrity of interpole connectors
	Check bolt torques, tightness on damper bar interconnections	Confirms mechanical integrity of damper bar interconnections

Note: Table 4-1 does not include excitation systems, bearings, or unit auxiliaries. These should also be inspected during a minor outage.

the generator. After the initial time period, and if the problems discovered during the annual inspection are minor, IEEE 492 recommends that minor scheduled inspection outages be performed every two or more years.[1]

The Rotating Machines Section, Group 11, of the Conférence Internationale des Grands Résaux Electriques (CIGRE) recommends minor periodic inspections every 30 months.[4]

Major Scheduled Inspection Outages

These inspection outages correspond to the IEEE's "complete inspections."[1] They include all the elements of a minor inspection as well as additional electrical and mechanical tests. Major inspections also include an extensive detailed visual inspection of the major and minor generator component parts. A major inspection outage normally includes removal of the rotor to allow a complete visual examination of the entire rotor and stator. The rotor removal requires an extended unit outage, varying from four to eight weeks.

Electrical tests typically included in a major inspection of a 100-MVA generator are listed in Table 4-2. A visual inspection of the stator and rotor during a major inspection includes a number of elements not included in the minor inspection (see sidebar, page 4-6). The major inspection also should include excitation systems, bearings, and unit auxiliaries.

Electrical tests are used to identify generator component part problems (windings, core, etc.) that are not evident visually. Although some of these tests are fail/not fail tests, many others provide data which should be compared to

TABLE 4-2: ***Additional electrical tests to be performed on a typical 100-MVA generator during a major inspection outage.***

Component	Test	Purpose of Test
Stator	Winding resistance	Detects shorted or defective winding circuits
	AC or DC high-potential test	Confirms insulation can sustain ratedvoltage without breaking down
	Power factor, dissipation factor, tan delta test	Confirms insulation dielectric loss is normal
	Core compression bolt insulation test	Confirms bolt is insulated from stator core steel
	Core interlaminar insulation test	Confirms integrity of stator core lamination and checks for core hot spots
Rotor	Rotor impedance test (with machine accelerating and decelerating)	Detects speed-dependent shorted turns on the rotor field windings
	Open-circuit test	Confirms open circuit characteristic of generator. Used to detect shorted turns in field windings
	DC high-potential test	Detects serious flaws in ground wall insulation

Chapter 4

Looking Deeper: Visual Inspections During Major Scheduled Inspection Outages

Removal of the rotor allows the inspector unimpeded access to the full length of the stator winding, stator core, and rotor including pole faces, damper connections, and field winding connections. This provides the opportunity to perform a detailed check for the following items:

Stator Windings

- ☐ Wedge tightness or migration
- ☐ Winding movement, fretting, abrasion, or gouges
- ☐ Cracks or tape separation in the winding insulation
- ☐ Slot or partial surface discharge
- ☐ Winding slot filler movement
- ☐ Insulation bulging into air ducts
- ☐ Rubbing by the rotor
- ☐ External discoloration of the insulation
- ☐ Loose end winding or associated bracing
- ☐ Circumferential/collector ring bus:
  - -overall integrity of insulation
  - -insulation wear adjacent to metal supports
  - -integrity of mechanical support to stator frame

Stator Core

- ☐ Bent or broken laminations
- ☐ Discolored or melted laminations
- ☐ Debris in air ducts
- ☐ Evidence of movement or fretting
- ☐ Rubbing by the rotor
- ☐ Excessive corrosion
- ☐ Compression tightness and clamping teeth
- ☐ Compression bolt insulation
- ☐ Waviness
- ☐ Stress fractures, rigidity in mechanical connections to frame

Rotor

- ☐ Evidence of heating on rotor pole face
- ☐ Damper bar interconnections from pole to pole
- ☐ Rotor fastenings, including dovetail keys or wedges that fasten the pole to the rotor rim
- ☐ Rotor rim
- ☐ Tightness or cracks on interpole V-block support
- ☐ Cracks or stress fractures on fan blades
- ☐ Field winding insulation over entire length of pole
- ☐ Cracks, looseness, broken welds, or evidence of movement in mechanical bracing or support components (hub, spider, and rim)

the factory test data, or field test data obtained when the generator was initially installed. If these tests reveal a problem, further testing and detailed inspections of the suspect component parts should be performed during the scheduled outage to confirm the specific problems.

Periodic maintenance tasks, such as changing brushes, lubrication, and adjustment of bearings, are usually performed at the same time as the inspection outage. Performing maintenance tasks in parallel with inspection activities minimizes the unit downtime required.

Determining the Frequency of Major Inspection Outages

Although there are "rules of thumb" for determining the frequency of major inspection outages, actual practice in the hydro industry varies widely.[5,6,7] Table 4-3, which shows the schedules of inspection outages followed by seven hydropower producers, illustrates this variation. Although these owners did not distinguish between major and minor outages, scheduled inspection outages within three years of each other are minor inspection outages unless otherwise noted.[8]

When Do You Need a Major Inspection Outage?

In general, major inspection outages should take place every five years. Owners may base the frequency of major inspection outages on factors such as the type and manufacturer of the generator, the owner's experience with the machine, and the conditions under which the generator is operated.

Certain characteristics of a generator indicate that it is subject to relatively low electrical and mechanical stresses, and may justify extending the period between major inspection outages. These characteristics include:

- Low speed (less than 200 rpm);
- Small physical size (rotor diameter less than 5 meters);
- Rated output less than 50 MVA;
- Terminal voltage less than 13.8 kV; and
- Part-time operation (due, for example, to flow restrictions or other environmental constraints).

Larger generators with higher operating speeds require more frequent major inspections. As a general rule, a major inspection outage should be conducted on any generator greater than 20 MVA with a terminal voltage higher than 6.9 kV at least two to three times during the generator's life.

CIGRE recommends that major periodic inspections be performed every 60 months.[4] IEEE Publication 492 does not identify a recommended time interval for conducting major scheduled inspection outages. This is probably due to the great diversity of generator designs and the various operating regimes under which they function.

TABLE 4-3: ***Frequency of Inspection Outages: Industry Examples***

Hydropower Producer	No. of Plants	Total MW	Scheduled Inspection Outage Frequency
American Electric Power	11	825	Every 18 months
British Columbia Hydro	29	9,706	Every two years
Duke Power	27	2,699	Based on running hours (did not specify number)
Tennessee Valley Authority	30	4,358	Every three years
Union Electric Company	3	681	Annually to five years
U.S. Army Corps of Engineers Districts:			
Great Lakes	2	21	Every five years
Kansas City	2	230	Annually
Omaha	6	2,454	Annually
South Atlantic	13	2,470	Every two years (major outages)
Southwest	18	1,725	Annually
Vicksburg	4	197	Annually
U.S. Bureau of Reclamation	58	14,311	Based on running hours (did not specify number)

IEEE Publication 492 does, however, state that "many manufacturers and users believe that major rotating electrical machinery should be completely dismantled and thoroughly inspected near the end of the first year of service." Although this is not a common industry practice, we have found that for large generators (200 MVA or larger) or highly stressed machines such as generator/ motors utilized at pumped storage plants, a major inspection outage after the first full year of operation can reap long-term dividends. The one-year inspection allows the owner to address problems in their preliminary stages during the initial break-in of the generator. The problems corrected during the one-year inspection could have resulted in forced outages three to four years later, before the first scheduled major outage. Loose stator cores and loose stator coil end wedges are among the problems commonly discovered on large generators after the first year of operation.

Using Predictive Maintenance Components to Determine Inspection Frequency

The increasing use and availability of predictive maintenance (PM) components can help owners develop appropriate inspection schedules for specific generators. However, the owner should recognize the fact that PM devices cannot replace a

FIGURE 4-1: ***Major Scheduled Inspection.*** *In a major scheduled inspection outage, the rotor is removed, the stator and rotor receive a thorough visual inspection, and individual generator components are subjected to an extensive series of electrical and mechanical tests.(Photo courtesy of U.S. Bureau of Reclamation)*

---

periodic inspection outage of the generators. This is because even the best PM devices marketed today cannot provide the quantity of detailed information on the generator internal components that is obtained by visual inspections and the associated electrical and mechanical tests performed during an inspection outage. The owner can best take advantage of PM devices by using their output in determining when and how frequently the minor and major inspection outages should take place.

Many generators, especially those installed prior to the 1980s, are not equipped with PM devices. Most of these generators will not be retrofitted with PM devices in the future, due to their small MVA size and the associated cost of installing the PM equipment. Generators without PM devices are totally dependent upon minor and major inspection outages to confirm their operational integrity.

Working with the Manufacturer

During a major scheduled outage, the generator manufacturer should be at the site to assist in examination of the generator. The manufacturer's assistance is important for several reasons. First, all generators incorporate unique proprietary features. These features have been developed throughout the manufacturer's history to strengthen the performance and reliability of the machine. The manufacturer's representative has specialized knowledge of the design and function of these features. The representative's general knowledge and experience in inspecting generators is also very helpful.

In addition, each manufacturer maintains a historical database of its generators that includes past inspection problems and deficiencies that caused major outages. The manufacturer uses this database to make design improvements. The historical database is invaluable during major outage inspections.

Planning Inspection Outages: A Common-Sense Approach

There clearly is no standard for establishing a program of inspection outages, nor is there a consistent practice within the hydro industry. The author has reviewed owner inspection outage programs that adhere rigidly to the manufacturer's recommended 5-year period for major inspection outages and programs that have not included a major scheduled inspection outage with a rotor removal since the generator was installed in the late 1930s.

When inspection outage programs are implemented, safety procedures and controls must be incorporated in the program to ensure that no tools or foreign objects are left inside the generator when the inspection is finished. Many costly generator failures have resulted from inadequate inventory and clean-out procedures.

In today's cost-competitive economic environment, a forced outage for repairs can be disastrous. This is especially true for small, financially vulnerable power producers. With this in mind, owners should use all the tools at their disposal, including manufacturer's recommendations, previous inspection results, common sense, and experience, to dictate a practical schedule of major and minor inspection outages.

REFERENCES

[1] "IEEE Guide for Operation and Maintenance of Hydro Generators," Publication 492, IEEE, 1976 (reaffirmed 1986).

[2] Nordrik, E. "Upgrading of Hydro Generators," *ELECTRA,* No. 162, October 1995.

[3] Electric Power Research Institute, *Power Plant Electrical Reference Series*, ISBN No. 0-8033-0002-6, Vol. 16, Chapter 6, 1989.

[4] Mitchell, A.S., "Rotating Machines," *ELECTRA,* No. 158, February 1995.

[5] Kerszenbaum, I., *Inspection of Large Synchronous Machines*, ISBN 0-7803-1148 5, IEEE Press, New York, NY, 1996.

[6] Maujean, J.-M., "EDF's Experience in the Repair and Maintenance of Hydro Generators," *Water Power & Dam Construction,* October 1989.

[7] Uberhorst, D., Weiland, H., and Wöhrle, G., "Life-Management Experiences with Generators in German Utilities," CIGRE Technical Paper 11-205, 1996.

[8] "Report on Hydro Operations Best Practices," HCI Publications, 1996.

Chapter 5
Transformers at Hydroelectric Plants

Electrical machines and transmission system components are designed to operate at specific voltage levels. The power transformer's function is to convert electrical power from one voltage level to another, permitting the power to be transmitted between system components operating at different voltage levels.

At hydroelectric plants, large transformers perform the primary task of delivering power produced by the generators to the transmission system. Other, smaller transformers serve the power needs within the plant itself. This chapter introduces the types and applications of transformers used at hydroelectric plants. Specific aspects of selecting, maintaining, and inspecting transformers are covered in the followng four chapters.

Transformers fall into two categories: liquid-immersed transformers, which are normally used for power transfer rates greater than 10 MVA and voltages higher than 34.5 kV; and dry transformers, which are normally used for power transfer rates less than or equal to 10 MVA and voltages below 34.5 kV. Liquid-immersed transformers also may be used for power transfer rates less than 10 MVA and voltages below 34.5 kV, if they are located outdoors. Liquid-immersed transformers are seldom utilized indoors because of the associated fire hazard.

The manufacture and testing of the two types are governed by separate standards, because of the great differences between the two in design and components.

Generator Step-Up Transformers

The power transformer that transmits the power produced by a hydroelectric plant to the utility network is commonly called the generator step-up (GSU) or main power transformer (MPT). In order to perform this task, the transformer must convert the low voltage at which the generator produces power (typically

4.16 to 24 kV) to a level that matches the transmission system (typically 69 to 500 kV).

The GSU transformer is a *liquid-immersed* transformer, meaning that the core and coils are in an oil-filled tank. The oil is circulated to external radiators to cool the core and coils during operation. Oil is a much more effective cooling and dielectric medium than air, and therefore is used for all transformers with capacities over 10 MVA.

The American National Standards Institute (ANSI) classifies GSU transformers separately from other liquid-immersed transformers.[1] GSU transformers are built with additional margins to allow them to operate as step-up transformers at a reduced power factor and 105 percent rated primary voltage on a continuous basis.

GSU Arrangements

The most common GSU transformer arrangements for a hydroelectric plant are illustrated in Figure 5-1. Scheme 1, the most common arrangement, is used at about 75 percent of hydro plants worldwide.[1] Although it is the most expensive of the three arrangements, this scheme is widely preferred for its simple low-voltage connections, reduced transformer sizes in comparison to alternative schemes, and unitized arrangement. In this arrangement, maintenance on a GSU transformer affects only one generator.

In Schemes 2 and 3, each GSU is shared between two generating units. This arrangement is used when initial cost savings are imperative or when powerhouse

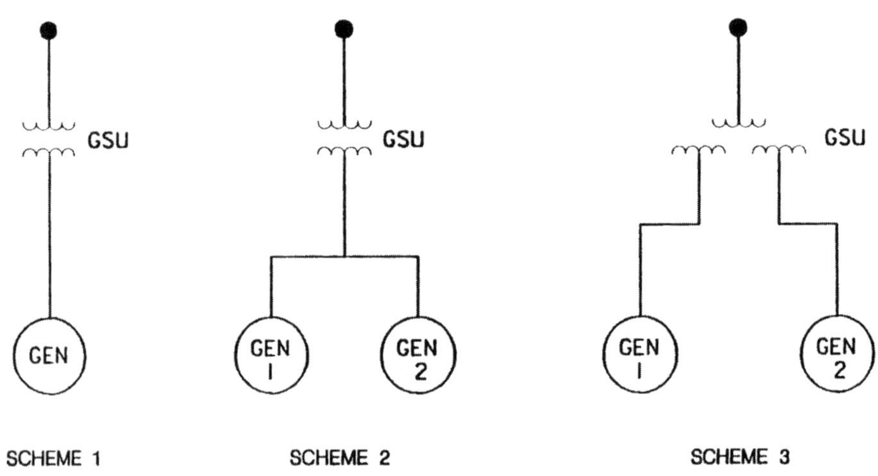

FIGURE 5-1: *GSU Transformer Arrangements.* *Although the Scheme 1 arrangement is widely preferred because of its simplicity and one-to-one pairing of generators and transformers, Schemes 2 and 3 may be selected to minimize initial cost or conserve powerhouse space.*

space is limited. However, there are two major disadvantages to both of these schemes: first, the loss of one GSU transformer results in the loss of two generating units; and second, replacement of a three-winding transformer is more costly than replacement of one GSU transformer in Scheme 1. (At some projects, three single-phase transformers have been used instead of a three-phase transformer; however, the three single-phase transformers have a higher initial cost than the three-phase transformer, require more space, and have a more complicated LV buswork than one three-phase transformer.)

In Scheme 2, the transformer's low-voltage coils are increased in physical size to accommodate the power output of the generator. In Scheme 3, the transformer is wound with two independent low-voltage coils feeding a single coil on the high-voltage side of the transformer. For Scheme 2, the fault current at the low-voltage terminals of the GSU is much greater than in Scheme 1 or 3. This must be accommodated in the associated switchgear and main leads design of the plant.

In certain countries, the history of hydropower development has led to a predominance of one scheme over the others. For example, Scheme 2 is used in many Canadian hydro plants and in about half of the Russian plants. Scheme 3 has been used extensively in Australia, less extensively in Italy, and occasionally in the United Kingdom, Sweden, Japan, and Norway. [2] Scheme 3 is not normally used in the United States.

Auxiliary Transformers

Transformers are also used at hydroelectric plants in the plant electrical auxiliary systems, which typically require from one to six percent of the unit MVA rating. Electrical auxiliaries include unit auxiliaries (systems, such as the excitation system or governor oil pressure system, that are directly associated with individual units) and station auxiliaries (systems associated with operating the station as a whole, such as a sump pump drainage system).

Although, in the past, different types of transformers have been used for serving the various plant auxiliary requirements, advances in dry-type transformer design have eliminated the need for liquid-immersed station auxiliary transformers. Transformers used for plant electrical auxiliaries today are usually dry-type transformers. Dry type transformers are air-cooled. They are manufactured with voltage ratings up to 34.5 kV and power capacities up to 10 MVA. Because air-cooled transformers present less of a fire hazard than liquid-immersed types, they are preferred for indoor applications at hydroelectric stations.

Auxiliary Transformer Arrangements

Power can be furnished to the hydro plant's electrical auxiliaries by the station auxiliary transformer (SAT) for station auxiliaries or by the unit auxiliary transformers (UAT) for unit auxiliaries. The SAT was also traditionally used to supply power during a unit start-up until the generator circuit breaker in the switchyard is closed, thereby connecting the generator to the network. However, with the introduction in the early 1980s of SF6 generator circuit breakers, which can be located inside the powerhouse in close proximity to the generator, there is no need to use station auxiliary transformers for start-up or station auxiliary power.

The two most common auxiliary schemes utilized at hydro power stations are shown in Figure 5-2.

Scheme 2 was the most commonly utilized scheme at projects predating the installation of SF6 generator circuit breakers in the powerhouse. Today, Scheme 1 is widely preferred for several reasons:

- It is more economical than Scheme 2, because it eliminates the SAT and the associated high-voltage connections. (The SAT used in Scheme 2 is usually an outdoor, liquid-immersed transformer.)
- It requires no complicated switchover of unit auxiliary power from the SAT to the UAT once the unit is synchronized and connected to the grid.
- It allows synchronizing on the low-voltage side of the GSU, thereby eliminating the high-voltage potential transformers and their associated cabling in the switchyard.

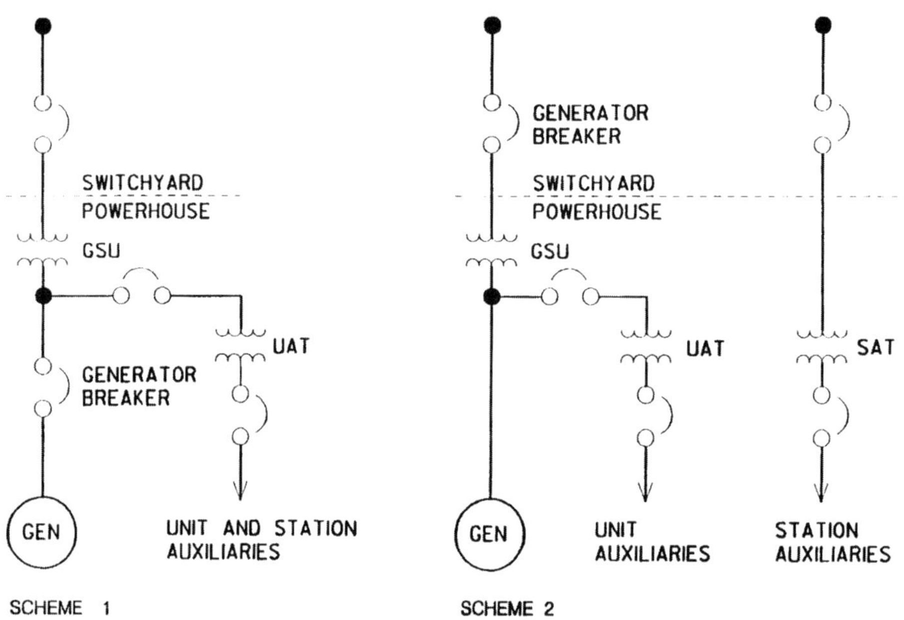

FIGURE 5-2: *Common Auxiliary Transformer Arrangements.* *This figure shows the two most common schemes for providing power to station and unit auxiliaries. Following advances in circuit breaker technology in the early 1980s, the more economical Scheme 1 became the method of choice.*

• It provides better protection for the GSU transformer, generator, and UAT in the event of a generator fault due to the fast clearance of the fault current.

The disadvantages of Scheme 1 are that it does not allow the plant to return to service if the UAT fails; and the GSU must remain energized whenever power is furnished to the plant auxiliaries. Therefore, the GSU must remain energized to supply the unit and station auxiliaries even during an extended maintenance outage. The GSU transformer losses during an extended outage may be significant, and should be evaluated. However, most owners find that the advantages of Scheme 1 outweigh its disadvantages.

Types of Auxiliary Transformers

Voltages on hydroelectric plant auxiliary power systems range from 20 kV medium-voltage systems to the normal 480-208/120 V low-voltage system. The SAT transformers are required to be liquid-immersed, which commonly means oil filled, due to the high voltages of their primary terminals. The primary terminals are connected to the high-voltage busbars of the substation. The SAT transformers must be located outdoors to minimize the fire hazard associated with a short-circuit condition or tank rupture.

Because the UAT is normally located inside the powerhouse near the auxiliary loads, dry type transformers are used for UATs. Dry type transformers reduce the fire hazard present with liquid-filled transformers (see Chapter 9). Two main types of dry-type transformers are used for UAT transformers: vacuum pressure impregnated (VPI) and cast coil epoxy resin. A third type, vacuum pressure encapsulated (VPE) transformers, are not normally used for UAT transformers, but are still used for smaller kVA transformers. VPI transformers normally are manufactured in sizes up to 6,000 kVA with primary voltage ratings up to 20 kV. Cast coil transformers can be made in sizes from 75 to 10,000 kVA, with primary voltage ratings up to and including 34.5 kV.

Although the same transformer theory applies to dry transformers as to liquid-filled units, the cooling medium, which is air, requires different temperature rise ratings, dielectric tests, and construction techniques to ensure reliability. Consequently, the factory and field tests for dry type transformers are established by a separate set of standards (ANSI/Institute of Electrical and Electronics Engineers (IEEE) Standards C57.12.91 and C57.12.01) from those utilized for liquid filled transformers.[3,4]

Purchasing Transformers for a Hydroelectric Plant

Writing Specifications

When purchasing transformers for a hydroelectric plant, the plant owner and engineers should be familiar with the factory testing methods and standards that will ensure the transformers' reliability and performance. (These tests are covered in detail in Chapter 6.) The specifications for the transformers should include tests to be performed, specific test procedures, and test values. The applicable standards typically allow wide degrees of latitude that may require engineering decisions at the factory. Therefore, the project owner should ensure that factory testing is witnessed by an engineer or other representative who has a sound understanding of the tests to be performed and the reasons for performing them.

Since a GSU transformer must conform to different ANSI standards than other liquid-immersed transformers, any specifications for a GSU transformer must state explicitly that the desired transformer is a GSU type.

Understanding International Standards

The market for supplying GSU transformers is international. North American hydroelectric plant owners can purchase high-quality equipment from domestic manufacturers or from firms based in Europe, South America, or Asia. While the location of manufacture does not necessarily affect product quality, important differences may exist in the standards applied to the equipment. Two sets of standards are preeminent internationally: ANSI and the International Electrotechnical Commission (IEC). Most North American and South Amercian manufacturers apply the ANSI code, while European and Asian manufacturers use the IEC standards. Though there are similarities between the requirements, their differences can have a substantial effect on both the buyer and the seller of large power transformers. Therefore, the equipment specifications should define which set of standards will be used. Chapter 8 addresses both systems of standards and includes examples of the differences between them.

References

[1] "General Requirements for Liquid-Immersed Distribution, Power, and Regulating Transformers," ANSI/IEEE Standard C57.12.00, American National Standards Institute(ANSI)/Institute of Electrical and Electronics Engineers (IEEE), 1987.

[2] Taylor, J., and D.J. Bornebroek, "Main Transformer Arrangements and Related Matters in Generating Stations," *ELECTRA*, No. 82, August 1981.

[3] "IEEE Standard Test Code For Dry Type Distribution and Power Transformers," ANSI/IEEE Standard C57.12.91, ANSI/IEEE, 1979.

[4] "IEEE Standard General Requirements For Dry Type Distribution and Power Transformers Including Those With Solid Cast and/or Resin Encapsulated Windings," ANSI/IEEE Standard C57.12.01, ANSI/IEEE, 1979.

Chapter 6
Factory and Field Testing of Liquid-Immersed Transformers

One essential component of a hydroelectric plant is the main power transformer (MPT), which converts the electricity produced by the generator from the generator's voltage to the voltage required by the transmission system. Most hydroelectric plants use an outdoor, oil-immersed MPT that can achieve a conversion efficiency greater than 99.5 percent, along with very high reliability and availability. Still, only careful attention to tests performed in the factory and at the site will ensure that the owner will obtain high reliability.

The major cause of transformer problems is insulation breakdown caused by such factors as transient voltage surges or partial discharges. Because insulation breakdown will eventually lead to a short-circuit failure, factory and field dielectric tests to verify the integrity of the insulation are essential to ensure transformer reliability. The dielectric tests are described in the American National Standards Institute (ANSI)/Institute of Electrical Engineering and Electronics (IEEE) standards, as are additional tests to confirm the transformer losses, impedance, turns ratio, winding resistance, temperature rise, and short circuit withstand capability. This chapter focuses on dielectric testing that should be performed both in the factory and on site for a liquid-immersed Class II transformer. (A Class II transformer has high-voltage windings rated 115 kV through 765 kV. ANSI requires less extensive dielectric testing for Class I transformers, which have high-voltage windings rated less than 115 kV).

Manufacture and testing of liquid-immersed transformers fall under ANSI/IEEE Standards C57.12.00 and C57.12.90, respectively.[1,2]

Chapter 6

Writing Test Specifications

Transformer specifications must contain a very complete, strongly phrased test section that explicitly lists the following:

- Required factory and on-site tests;
- Testing procedures to be followed in the event that the ANSI/IEEE

FIGURE 6-1: ***Testing Transformers.*** *The main power transformer for the 12-MW Moose River Hydroelectric Project in New York State was installed in 1987. A full spectrum of dielectric tests was performed on the 16.6-MVA transformer at the time of installation. As of early 1998, there had been no transformer-related outages at the plant since the installation.*

standard allows different test methods (if the owner or engineer has preferences); and

- Specific test values to be obtained in tests for which ANSI/IEEE does not specify standard values.

The specification also should state that a qualified representative of the owner will witness all tests.

Because there are always decisions that must be made during testing, the witness should clearly understand not only the mechanics of the test, but also the rationale behind each test. Typical decisions include the acceptability of the impulse wave shape generated by the manufacturer's surge generator; the position of the tap-changer(s) during the various dielectric tests; and the method to be used to test the high-voltage neutral terminal (ANSI/IEEE C57.12.90 provides two alternative methods).

For all tests performed in both factory and field, the manufacturer must complete certificates documenting the test procedures and the results achieved.

FACTORY DIELECTRIC TESTS

ANSI/IEEE C57.12.00 includes several dielectric tests for Class II transformers, as listed in Table 6-1. However, the standard states that the "routine" tests for a

Class II transformer consist of only the low frequency tests and the impulse test.[1] Consequently, the other tests listed in Table 6-1 must be explicitly required by the specifications; otherwise the manufacturer is not required to perform these tests on the transformer at the factory. Further, neutral terminals are not impulse tested unless the specifications so stipulate. Impulse testing of the neutral terminals is a matter of judgment; some conservative utilities require this test for all transformers.

In the author's opinion, insulation power factor tests and radio influence voltage/partial discharge tests should be performed on all main power transformers 15 MVA or larger that have a high-voltage winding rated at 115 kV or greater. In addition, the types of impulse tests performed on the transformer—reduced full wave (RFW), front-of-wave (FOW), chopped wave (CW), or full wave (FW) —should be included in the specifications to preclude any arguments or misunderstandings during the factory tests. The switching impulse test is usually appropriate for network transformers rather than main power transformers.

TABLE 6-1: ***Dielectric Tests for Class II Transformers (501 kVA and Larger)***

Routine tests (performed on all transformers)

Low frequency tests: applied voltage and induced voltage

Lightning impulse tests

Design Tests (performed on a specified number of newly designed or modified transformers)

None

Other Tests (should be required by owner's specifications in addition to routine tests)

Front-of-wave impulse test

Switching impulse, phase-to-ground

Radio influence voltage/partial discharge

Insulation power factor

Insulation resistance

-Adapted from ANSI/IEEE Standard C57.12.00, Table 16.[1]

Applied Voltage Test

The applied voltage test stresses the transformer's phase-to-ground insulation (the major insulation between the windings and the tank). ANSI/IEEE C57.12.00 specifies the test voltage, which is applied at 60 Hz for one minute.[1] Initially, 25 percent of the required voltage is applied; this voltage is gradually increased to the required value before de-energizing the test circuit. To prevent the accidental application of excessive over-voltage to the transformer, the test circuit includes a relief or spark-over gap set approximately 10 percent higher than the required test voltage.

Chapter 6

The test procedure consists of interconnecting the three terminals being tested (for example, the high-voltage terminals), and connecting these terminals to the voltage generator while the terminals of the other three (low-voltage) windings are jumpered together and connected to the tank, which is grounded.

This procedure is modified for delta-grounded wye (low-voltage/high-voltage respectively) transformers on which the high-voltage-neutral bushing has a reduced or graded insulation value with respect to the other high-voltage terminals. ANSI/IEEE C57.12.00 allows all high voltage terminals for Class II power transformers, including the neutral, to be tested at the neutral insulation level for wye-winding grounded transformers.[1] This is a significantly lower test voltage than for Class II ungrounded transformer windings. The rationale for this reduced voltage level is that the neutral voltage shift for a grounded winding will not exceed the lower test voltage even during a short-circuit condition.

An additional item to be verified for this test is the setting of the high-voltage tapchanger. The tapchanger should be set at its minimum value in order to place the maximum dielectric stress upon the high-voltage winding.

Induced Voltage Test

The induced voltage test stresses the transformer's turn-to-turn and phase-to-phase insulation at an "enhancement voltage level" for one minute and at a reduced voltage level for one hour. The test voltage, which is selected from Table 5 of ANSI/IEEE C57.12.00, normally is a single-phase voltage applied to the low-voltage terminals.[1] This voltage induces the required test voltage on the high-voltage terminals, which remain open-circuited. Figure 6-2 shows a typical single-phase test arrangement.

The test is performed at a higher frequency than 60 Hz because the per-unit voltage required to test a three-phase transformer would over-excite the transformer core. This is based on the equation

$$E = 4.44 f \phi N$$

Because transformer cores are designed for rated primary voltage (E) at 60 Hz (f), the turns ratio (N) and the voltage (E) are fixed, leaving the flux density (ϕ) as the only variable in the equation. To fabricate their transformers as economically as possible, manufacturers attempt to obtain the required flux density using a minimum amount of core steel. To maintain the flux density within the original design parameters, the higher-than-rated voltage required for this test is produced by doubling the frequency component (f) to 120 Hz.

The duration of the enhancement voltage test is 7,200 cycles, which means that at a frequency of 120 Hz, the test should last one minute. The duration of the reduced voltage level test is one hour. During the one-hour test, partial discharge levels are monitored, and must not exceed 100 microvolts. In general, the standard procedure is to maintain excitation current below 30 percent of the rated load current to avoid over-excitation of the transformer core.[4,5]

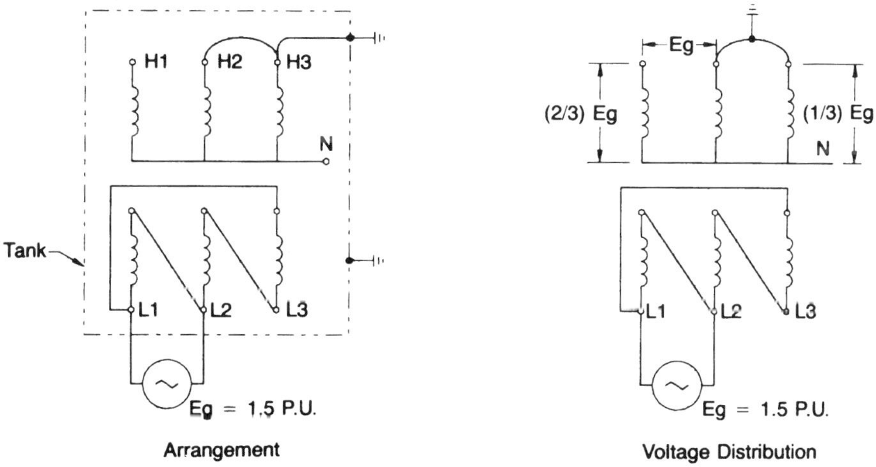

FIGURE 6-2: ***Induced Voltage Test.*** *The arrangement for an induced voltage test, which is a single-phase test of high-voltage/low voltage windings on the transformer. Eg is the voltage generator; H1, H2, and H3 are high-voltage terminals, and L1, L2, and L3 are low-voltage terminals.*

Impulse Tests

Impulse tests simulate lightning surges and flashovers (either line-insulator flashovers or surge-arrester flashovers) on a high-voltage transmission line. These tests primarily stress the front end of the windings, winding connections to the bushing, and the bushings themselves. Over-voltages between these various parts may exceed 100 to 300 times normal values unless the transformer is properly designed.[4] There are several types of impulse tests, simulating different over-voltage situations, which are conducted in sequence. The impulse test sequence is specified by ANSI/IEEE C57.12.90.[2] (Internationally, the required impulse test sequence varies from the ANSI/IEEE standard. Differences between ANSI and IEC standards are detailed in Chapter 8.)

Types of Impulse Tests. The surge that would be caused by a lightning strike on the transmission system some miles away from the transformer is simulated by the *full wave impulse* of 1.2 x 50 microseconds (1.2 microseconds for the voltage to rise to its peak value; 50 microseconds for the voltage to decay to 50 percent of the peak value). The required peak voltage depends on the voltage class of the transformer. Under ANSI standards, a *reduced full wave* test, with the same shape but peaking at either 80 percent or 62.5 percent of the full wave peak, is applied to the transformer before the standard full wave test.

The *chopped wave* test is more severe than the full wave test. It simulates a surge arrestor firing, or insulator flashover caused by a lightning strike in the immediate vicinity of the transformer. This condition is simulated by chopping, or

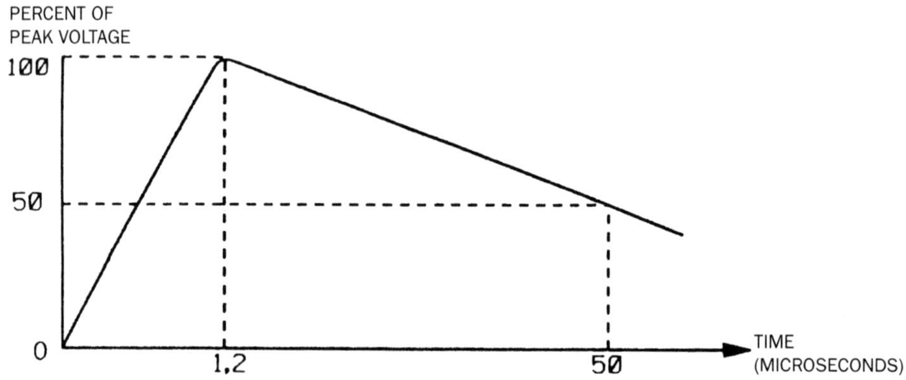

FIGURE 6-3: ***Full Wave Lightning Impulse Test.*** *This test simulates the surge that would be caused at the transformer by a lightning strike occurring elsewhere in the transmission system.*

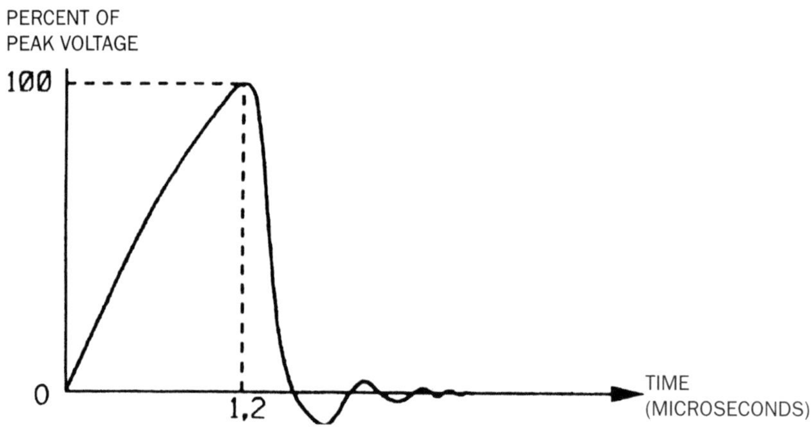

FIGURE 6-4: ***Chopped Wave Lightning Impulse Test.*** *This test simulates the surge that would be caused by a lightning strike close to the transformer.*

interrupting, the wave front, causing a rapid collapse of the voltage.

The *switching surge impulse test* simulates the operation of a circuit breaker or switching device on the transmission network to which the transformer is connected. This wave has the same general shape as the full wave, except that the times to the peak voltage and the 50-percent-of-peak point are longer (100 microseconds and 1,000 microseconds, respectively).

The front-of-wave impulse test is an additional test that can be specified if there is no direct lightning protection provided for the transformer, such as substation shield wires. This test is more stressful to the front end of the transformer

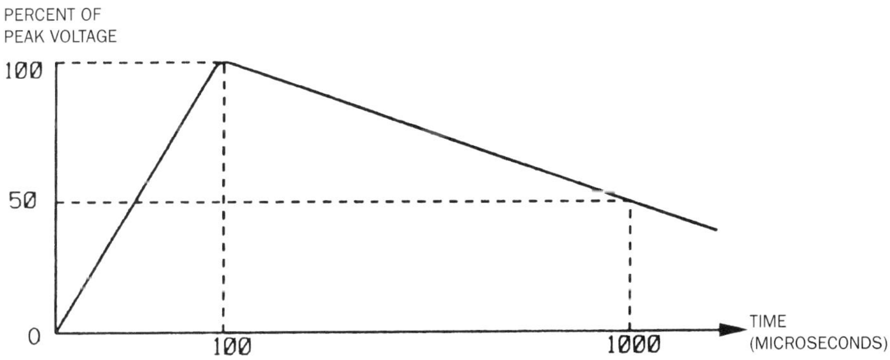

FIGURE 6-5: ***Switching Surge Impulse Waveshape.*** *This test simulates the surge that would be caused by operation of a circuit breaker or switching device on the transmission network.*

FIGURE 6-6: ***Lightning Impulse Testing.*** *In this photograph, 400-MVA generator step-up transformers for the 760-MW Rocky Mountain pumped-storage project in Georgia are being given lightning impulse tests at the Siemens Trafo-union GmbH plant in Germany.*

than the chopped wave test, but is rarely required because engineers normally provide direct lightning protection for the substation and connecting transmission lines.

Generating Waveforms for Testing. Because the generation of high-voltage impulse waves is not an exact science, it is not always easy to achieve the specified 1.2 x 50-microsecond wave shape when testing a large transformer. ANSI/IEEE tolerances are fairly wide because they recognize the complications involved in generating a specific wave shape. To obtain a wave shape within ANSI/IEEE tolerances, the manufacturer must experiment with adding different resistances and capacitances to the impulse-generator circuit. To minimize space requirements on the factory test floor and to eliminate inductance associated with connecting cables, all resistors associated with the wave-shaping circuit are incorporated in modern impulse generators.

Frequently, the manufacturer will make many adjustments to the impulse

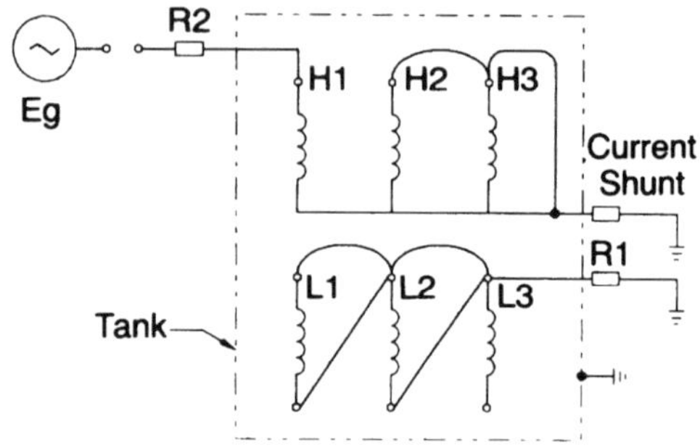

FIGURE 6-7: ***Impulse-Generator Circuit.*** *This arrangement tests one winding (H1) of a transformer. A current shunt between neutral and ground is used to convert neutral current waves to voltages. Eg is the voltage generator; R1 and R2 are wave-shaping circuit resistors; H1, H2, and H3 are high-voltage terminals; and L1, L2, and L3 are low-voltage terminals.*

generator circuit, finally producing two or three different wave shapes with different rise times and tails, all within ANSI/IEEE tolerances. At this point, the owner's representative must choose the most appropriate wave for testing. One good approach is to select the impulse wave that has the fastest rise time, as this is the most stressful test of the front end of each transformer winding. In general, the impulse wave shape produced by the impulse generator will be checked against the center winding of the transformer. Although all windings should have identical impedances, slight capacitance-to-ground differences may exist between the left and right windings. If the wave is initially tested against the center winding, this slight difference should not appreciably affect the impulse wave shape.

Each high-voltage and low-voltage terminal is impulse tested individually.

Before arriving at the factory to witness impulse tests, the owner's representative should consider two related factors: the position of the high-voltage and low-voltage tapchangers; and the monitoring of the neutral-current wave shape in addition to the terminal voltage wave shape.

Positioning the High-Voltage and Low-Voltage Tapchangers. To apply maximum dielectric stress to the high-voltage tapchanger of the winding undergoing impulse testing, the tapchanger should be set at its minimum tap, according to paragraph 10.3.2.4 of ANSI/IEEE C57.12.90.[2] Applying maximum stress will verify that the tapchanger should work reliably over the lifetime of the transformer. This is an important check, as the tapchanger and other accessories account for a high percentage of transformer failures. [6,7]

If the transformer has both high-voltage and low-voltage tapchangers, the high-voltage tapchanger should be set at its minimum and the low-voltage set at its maximum for impulse tests of the high-voltage terminal. For the low-voltage terminal impulse tests, the low-voltage tapchanger should be set at its minimum, and the high-voltage tapchanger at its maximum. These settings of the tapchangers will apply maximum dielectric stress to each tapchanger during one sequence of impulse testing.

Monitoring the Neutral-Current Wave Shape. Data to be recorded for each impulse test should include not only the terminal-voltage wave shapes, but also the neutral-current wave shapes, according to paragraph 10.3.4.1 of the ANSI/IEEE C57.12.90.[2] This is accomplished on a wye transformer by inserting a current shunt between the neutral point and ground, as shown in Figure 6-7. On a delta transformer, the current shunt is inserted between the two terminals not being tested (which are jumpered together) and the ground point. The neutral-current wave shapes provide additional comparative data for the reduced full wave (RFW) and full wave (FW) tests. Any irregularity that appears when reviewing the terminal-voltage oscilloscope photographs can be positively confirmed by comparing the neutral-current wave shapes.

To analyze the terminal-voltage and neutral-current wave shapes, the owner's representative makes a point-by-point comparison of the RFW and the FW tests for each transformer terminal tested. In general, the neutral-current photographs will provide more easily detectable evidence of a failure than will the terminal-voltage photographs. This is because a fault in the winding being tested will cause the inductance of the winding to vary, and consequently the inductive component of the current being monitored will also vary. Terminal-voltage oscillogram photographs show far less wave shape variation caused by a fault in the winding.

Faults that occur during the two chopped wave (CW) impulse tests are extremely hard to detect by photographic comparisons; however, they should be detectable when the final full wave test is performed. This is the reason the ANSI/IEEE standard sequence for each terminal is RFW, CW, CW, FW.

To initiate impulse arc-over and consequently obtain the impulse wave shape desired, most manufacturers' test facilities use a solid-state ignition device rather than the older ball-gap devices. Ball-gap devices arc-over at different voltages, depending upon air temperature and humidity. Accordingly, they produce inconsistencies in the impulse wave shape applied to the different transformer terminals. Solid-state ignition devices provide a more consistent impulse wave form that reveals failures more clearly in oscillogram photographs.

Insulation Power Factor/Dissipation Factor Test

This test measures the electrical energy dissipated as heat when the insulation is stressed with an electrical AC field. The main value of the insulation power factor/dissipation factor test is in detecting the presence of poor or deteriorated layers of insulation in series with good insulation layers.[8] This test is described in paragraph 10.9 of ANSI/IEEE C57.12.90.[2]

There are no perfect insulating materials; when stressed by an AC electric field, all insulating materials convert some electrical energy into heat energy. To prevent this heat from causing thermal deterioration and subsequent breakdown of the insulation, dielectric losses generally must be low and insulation resistance high.

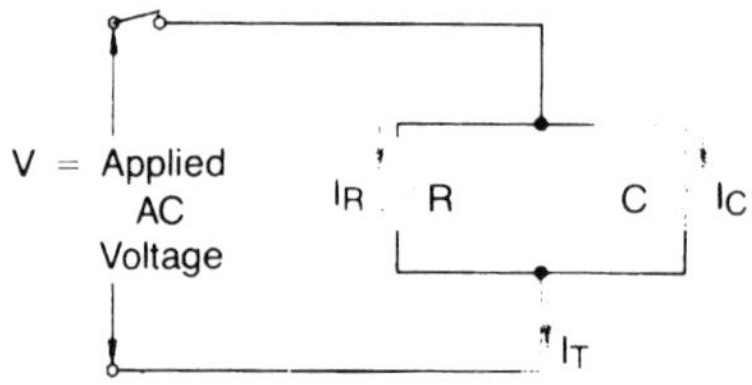

Equivalent Electric Circuit

R = Insulation Resistance
C = Insulation Capacitance
Dielectric Power Loss = $V I_R = I_R^2 R$ = Dissipation Factor
I_T = Total Charging Current
I_R = Leakage Current
I_C = Capacitive Charging Current

$$PF = \frac{\text{Dielectric Power Loss}}{\text{Total Power Applied}}$$

$$PF = \frac{V I_R}{V I_T} = \frac{I_R}{I_T}$$

FIGURE 6-8: ***Insulation Power Factor/Dissipation Factor Test.*** *In this test, an AC current is applied to the winding insulation and the charging current and leakage current vectors are recorded and compared.*

The power factor (PF) is defined as the ratio of the resistive (leakage) current measured in the insulation circuit to the total charging current. For a 10-kV/AC test set, insulation power factor values of 0.5 percent or less for each winding indicate excellent insulation.[5] The dissipation factor (DF), or tan delta, is the tangent of the angle between the capacitive current vector and the total current. The power factor/dissipation factor test is sometimes called the "Doble Test," from the name of the manufacturer of the test equipment.[9] In each case, the tester applies the specified AC voltage, and the equipment measures the resistive (leakage) and charging current vectors, which provide the PF value or the DF/tan delta value and indicate the dielectric power loss of the insulation.

The factory DF test is an excellent maintenance tool because the results can be compared with DF test values measured in the field later in the life of the transformer. Moreover, even though a DC high-potential test performed in the field may indicate adequate insulation, a DF test could show that losses have increased over those measured in previous years. This suggests that the insulation is breaking down because of partial discharges, and insulation replacement should be considered.

Insulation Resistance Tests

Commonly called the "megger" or "megohmmeter" test, the insulation resistance test stresses the insulation by applying from 500 to 1,000 DC volts, then measures the milli- or microamps DC leakage current that flows through the insulation. From this current reading and the known fixed voltage applied, the tester can determine the insulation's megohm resistance.

A high resistance value indicates that leakage current through the insulation is minimal, and the circuit may be safely energized. To be acceptable, the megohms measured on each winding (corrected to a specific temperature)

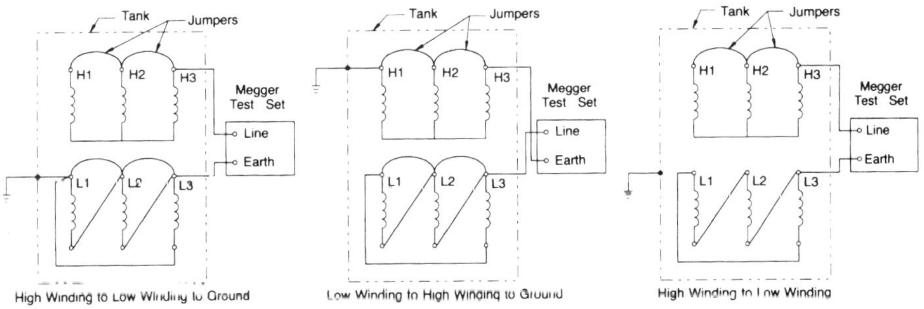

FIGURE 6-9: ***Insulation Resistance Test.*** *These schematics show the typical connections for performing the insulation resistance test on a delta-wye transformer for three different tests. H1, H2, and H3 are high-voltage terminals and L1, L2, and L3 are low-voltage terminals.*

should, at a minimum, be greater than the value of the rated kV + 1. This usually allows a leakage current of less than 1 milliampere, which is unlikely to damage the insulation.[10]

Two variations of the insulation resistance test are the *dielectric absorption test* and the *polarization index test.* In the dielectric absorption test, the DC voltage is applied for five to 15 minutes, and the leakage current is recorded periodically. If the insulation is in good condition, its resistance should increase with time; consequently, leakage current readings will decrease with time. The ratio of starting leakage current to ending leakage current should normally be greater than 1.25 in order to confirm insulation integrity.

The polarization index test measures the ratio of the insulation resistance value obtained after ten minutes of testing to the value obtained after one minute of testing. The minimum polarization index should be two or greater.

Partial Discharge Tests

Partial-discharge tests detect corona discharges within or on the insulation. Capable of weakening the insulation over a period of time by themselves, corona discharges also generally indicate irregularities or overstressing of the insulation system. Those developments can lead to insulation failure in the vicinity of the discharge.

Partial discharges can be measured using either narrow-band or broad-band methods. The partial discharge measuring equipment is usually connected to the transformer by means of special coupling capacitors and a measuring impedance. Calibrating the partial discharge circuitry is quite complicated. The measuring circuit must be tuned to the narrow-band frequencies required by ANSI to detect partial discharges in these ranges. Moreover, completely Faraday-shielded rooms must be used for accurate partial-discharge measurements to eliminate extraneous

Chapter 6

noise from other electrical equipment.

ANSI C12.57.90 stipulates that calibration must be conducted on each high-voltage terminal by injecting a known partial-discharge input (commonly called a PD charge) to the terminal while examining the response of the measuring circuits.[2] Several experts have identified the potential for errors concerning the measurement systems. These error tolerances are accounted for during the measuring circuit calibration procedure. Once the test circuit has been calibrated, no subsequent changes or adjustments may be made to the measuring circuit for the duration of the partial-discharge test.

ANSI specifies a detector with a narrow-band frequency response of 1 MHz to test for partial discharge. The advantage of narrow-band measurements is that external noise or electrical interference can be more easily eliminated by selecting a suitable frequency. In contrast, broad-band partial-discharge measurements have the advantage of greater sensitivity. Partial discharge is known to generate a wide spectrum of frequencies. However, there is not presently an internationally accepted range of test values for the broad-band method of measurement. Consequently, narrow-band is the prevalent method in the industry today.

To detect a partial discharge in the actual factory test, a tuned radio frequency circuit is used during a one-hour, 1.5 per unit (1.5 times the rated line-to-ground voltage) induced potential dielectric test. ANSI states that the transformer has passed the test if:

- Maximum partial discharge sensed is less than 100 microvolts;
- The increase in discharge levels does not exceed 30 microvolts; and
- A steady increase in discharge does not occur throughout the hour.[2]

During the one-hour test period, partial-discharge measurements with the calibrated circuit are made at five-minute intervals on each high-voltage terminal. Occasional high partial-discharge readings should be disregarded as they are probably due to extraneous electrical noise. If the partial discharge level rises above 100 microvolts for a certain time period and then returns below this level, the test must continue without interruption until below-maximum levels are obtained for a required period. The ANSI standards require acceptable levels for 60 minutes. If the transformer partial-discharge levels do not return to acceptable levels for the defined time period, the transformer is considered to have failed the test.

The ANSI code includes the general caveat: "A failure to meet the partial discharge acceptance criteria should not therefore warrant immediate rejection of the transformer, but should lead to consultation between purchaser and manufacturer about further investigations." This qualification was included because partial-discharge measurements are relative. This implies that the owner's representative witnessing the factory tests should have technical training and experience in partial-discharge testing in the event alternate decisions must be made concerning the test procedure or rejection of the transformer.

OTHER TRANSFORMER TESTS

Efficiency Tests

Transformers are very efficient electrical machines. The average efficiency for GSU transformers exceeds 99.7 percent. However, many GSU transformers are very large devices with ratings in excess of 100 MVA. As a result, even small decreases in efficiency are noteworthy, especially considering that the efficiency loss can continue for the 30-year life of the transformer. Efficiency testing at the factory is required by ANSI to confirm the losses, and inversely, the efficiency of the transformer.

Transformer loss measurements are defined in two categories: load losses and no-load losses. Load losses consist of the copper or winding losses of a transformer that occur when the transformer is transmitting power. No-load losses are the residual iron losses of a transformer. They occur any time the transformer is energized, whether it is transmitting power or not.

Mechanical Tests

In addition to electrical tests, transformers are subjected to a number of mechanical tests at the factory that are intended to ensure the integrity of the tank and auxiliary cooling equipment.

Leak Test. With the oil at normal operating temperature and level, the transformer tank and radiators should be pressurized to 125 percent of the maximum operating pressure for a period of 24 hours. Both during the test and at its conclusion, the inspector should check carefully to detect any oil or air leaks.

Pressure Test. The transformer tank and radiator should be tested at a vacuum of 0.05 inches of mercury absolute to verify that they can be vacuum-filled with oil in the field.

Weld Tests. The specifications should allow the owner or engineer to "spot check" tank welds in the factory, using dye-penetrant, ultrasonic, or other appropriate tests.

Specifying the Factory Test Sequence

In addition to specifying which tests are to be performed, the owner's representative also must consider the appropriate testing sequence. Paragraph 10.1.5.1 of ANSI/IEEE C57.12.90 stipulates that impulse tests shall precede the applied voltage, induced voltage, and switching surge tests. Furthermore, paragraph 4.4 of the standard states that the resistance, polarity, phase relation, impedance, no-load, load-loss, and temperature-rise tests should precede the dielectric tests.[2] This is the logical test sequence for large transformers. Because the tests preceding the dielectric tests are conducted at reduced voltages and currents, any failure that might occur would inflict minimal damage on the transformer. In addition, performing the temperature-rise test prior to the dielectric tests ensures that the insulation oil is uniformly distributed throughout the core, which will help avoid any breakdown in the core-steel laminations caused by spark-overs during the impulse tests.

Some owners prefer to have the no-load tests performed both before and after the impulse tests. If the results change more than 15 percent, there is sufficient basis for rejecting the transformer. A variation of this magnitude could indicate that a number of core-steel plates were welded together during the impulse tests.

The average increase of no-load (core) losses after the impulse tests ranges from 2 to 10 percent.[3] However, these increases seem to be decreasing in recent years as manufacturers improve their methods of cutting and assembling the core materials.

Field Tests

After factory testing has verified that the transformer meets the specifications, only a few on-site tests are required before the transformer is put into service at the hydro plant. The insulation resistance test and the insulation power factor tests are repeated to confirm the dielectric strength of the insulation, and a sample of the transformer oil is extracted and analyzed. Satisfactory completion of these tests will confirm that the transformer can be energized without problems.

Transformer tests should be performed by the manufacturer's field test engineer. This will insure that the method of testing is acceptable to the manufacturer, and will help to avoid warranty problems. Testing should be witnessed by an owner's representative who has the background and knowledge to understand the tests and to confirm that they have been performed properly.

It is important to compare the results of the field tests to the factory test results, and to investigate thoroughly any discrepancies. If discrepancies are acceptable, record the reasons on the field test report. These notes will be important when future field tests are performed.

Several other considerations are important during installation of the transformer.

Oil Analysis. File the laboratory oil analysis with the transformer certified reports. This oil analysis will give plant maintenance personnel a base line for comparison to analyses of oil samples extracted from the transformer in the future.

Grounding. Ensure that all ground connections to the transformer, surge arrestors, and auxiliary equipment are properly connected to the power plant ground grid. Check these ground connections for continuity.

Tank Positive Pressure. Ensure that positive pressure is maintained over the oil inside the transformer tank if the transformer is outside, unenergized, for any period of time. If internal tank pressure falls below atmospheric pressure, contaminants such as moisture and dust will penetrate into the transformer tank, degrading oil quality.

REFERENCES

[1] "General Requirements for Liquid-Immersed Distribution, Power, and Regulating Transformers," ANSI/IEEE C57.12.00, American National Standards Institute(ANSI)/Institute of Electrical and Electronics Engineers (IEEE), 1987.

[2] "Test Code for Liquid Immersed Distribution, Power, and Regulating Transformers and Guide for Short-Circuiting Testing of Distribution and Power Transformers," ANSI/IEEE C57.12.90, ANSI/IEEE, 1987.

[3] Kan, H., "Problems Related to Cores of Transformers and Reactors," *ELECTRA*, No. 94, May 1984.

[4] Manning, M.L., "Fundamentals of Insulating Transformers," *IEEE Electrical Insulation*, Vol. 3, No. 6, November 1987.

[5] Gill, A.S., *Electrical Equipment Testing and Maintenance*, Reston Publishing Company, Reston, Virginia, 1982, pages 270 and 276.

[6] "An International Survey on Failures of Large Transformers In Service," Final Report of Working Group 05 of Study Committee 12, *ELECTRA*, No. 88, May 1983.

[7] Allan, D.J. "General Report of Group 12, Transformers," *ELECTRA*, No. 85, December 1982.

[8] Seeberger, Rene, "Capacitance and Dissipation Factor Measurements," *IEEE Electrical Insulation*, Vol. 2, No. 1, January 1986.

[9] Meyers, S.D., Kelly, J.J., and Parish, R.H., *A Guide to Transformer Maintenance*, Main Transformer Institute, Division S.D. Meyers Inc., Akron, Ohio, 1981, page 400.

[10] Stone, G.C., and M. Kurz, "Interpretation of Megohmeter Tests on Electrical Apparatus and Circuits," *IEEE Electrical Insulation*, Vol. 2, No. 1, January 1, 1986.

BIBLIOGRAPHY

"IEEE Standard for Gapped Silicon-Carbide Surge Arrestors for AC Power Circuits," ANSI/IEEE Standard 62.1-1981, IEEE, 1981.

Chapter 7
Inspection and Maintenance Testing of Liquid-Immersed Transformers

Generator step-up (GSU) transformers at hydroelectric plants are oil-filled, liquid-immersed transformers. The oil acts as a cooling and insulating medium for the transformer. With appropriate inspection and maintenance procedures in place, the life expectancy of an outdoor GSU transformer is approximately 20 to 25 years, according to the American National Standards Institute (ANSI). The ANSI figure is based on an assumed 65-degree Celsius temperature rise at an ambient temperature of 40 degrees Celsius.[1] However, due to lower seasonal ambient temperatures and the fact that the transformer is not continuously operated at rated MVA, actual life spans of transformers are typically 30 to 40 years.

Maintenance testing for GSU transformers consists of a thorough visual inspection in conjunction with routine electrical tests on the windings and the oil. Separate electrical tests for the transformer bushings are also essential maintenance procedures.

Transformer Inspection and Maintenance Tests

Visual Inspection

An inspection of a transformer should begin with a complete visual inspection. The visual inspection should cover the following areas:

Transformer Tank. The tank should be thoroughly examined for corrosion or oil leaks. Although the inspection should cover the entire tank, the inspector should be aware of several particularly likely trouble spots. These include the gasketed areas, such as bushings, valves, and radiator flanges, and the lower section of the main tank.

Transformer Bushings. The external porcelain of the bushings should be checked for cracks or contamination. If there has been excessive leakage current, the porcelain surface will usually show visible evidence of this condition in the form of leakage current tracks. (A nominal amount of leakage current through

Chapter 7

the bushing insulator is normal and inevitable. However, if the bushing porcelain is damaged or heavily contaminated, the leakage will increase to an unacceptable level.)

Bushings are usually oil-filled for ratings above 34.5 kV to take advantage of oil's high dielectric strength relative to air. The oil level of the bushing should be checked, if possible, by viewing either an oil sight glass located at the top of the bushing or an oil level gauge or pointer. Historically, U.S. manufacturers provided oil sight glasses or gauges only for bushings with voltages of 69 kV or more. However, recently-manufactured bushings with ratings as low as 25 kV may be equipped with oil sight glasses.

Wiring Connections. The wiring in the main termination cabinet should be inspected for any loose connections. In addition, the inspection should verify that the cabinet is securely sealed to prevent rodents and insects from entering.

Thermometers and Level Indicators. The inspection should include a check of all dial type thermometers and level indicators to ensure that their readings are within recommended normal ranges. These include the liquid temperature indicator (oil thermometer), hot spot temperature indicator, liquid level indicator, and other indicators depending on the specific equipment.

Nitrogen System (if applicable). The nitrogen system is a positive-pressure, inert gas system that helps protect the transformer oil from atmospheric contamination. The various components of the nitrogen inert gas system (gauges, piping, nitrogen bottles, etc.) should be inspected to confirm that no problems are evident. Typical problems are excess gas use, gas leaks, and negative pressure in the main tank.

Conservator System (if applicable). The conservator system, in which a secondary oil tank is situated above the main transformer oil tank, also helps protect the transformer oil from contamination. The inspection should cover the gauges, conservator tank, Buchholz relay (gas detection system), and associated piping. The silica gel dryer should also be inspected to confirm that it is functional and does not need replacement. Usually the dryer should be replaced if one-third of its volume has changed color from blue to red.

Sudden Pressure Devices. The mechanical sudden pressure devices are located on the top of the tank. If an overpressure has occurred in the tank, triggering operation of the device, a mechanical target is usually visible. The sudden pressure device should also be checked for evidence of corrosion or leakage.

Auxiliary Fans, Pumps, Motors, Radiators, etc. These areas should be inspected for obvious problems such as bird nests, mechanical damage to the fans, oil leaks, or water leaks.

Ground Connections. The main tank and surge arrestor ground connections should be checked for significant corrosion or loose or broken connections.

Internal Inspection. In general, an internal inspection of a transformer should not be attempted unless indicated by the results of an electrical maintenance test (covered in the following section). Internal inspections are not normally performed in the field because of the high risk of introducing contaminants (moisture, dirt, and dust) into the transformer oil.

Routine Transformer Maintenance Tests

Electrical maintenance and infrared tests identify transformer problems that require immediate attention or correction. Oil tests are used to provide an early indication of possible future problems with the transformer.

Electrical Maintenance Tests. Electrical maintenance tests are performed on the high voltage/low voltage windings of the transformer. The most common electrical tests performed on a transformer are:

- Insulation resistance, or "megger," test;
- Power factor/tan delta test;
- Power factor tip-up test;
- Excitation current test; and
- Turns ratio test.

TABLE 7-1: *Electrical Maintenance Tests for GSU Transformers*

Test	Purpose	Acceptable Test Values
Insulation Resistance ("Megger Test")	Direct current test of insulation resistance	Minimum of 1 Megohm per kV of operating voltage of the transformer, or maximum of 1 milliamp per kV of leakage current
Power Factor/Tan Delta ("Doble Test")	Alternating current test of insulation resistance	Good: 0.5% or less Marginal: 0.5% to 1.0% Investigate: above 1.0% *(assuming 10 kV test set)*
Power Factor Tip-Up	Measures the difference between power factors measured at two different voltages	Any change from initial test values indicates change in insulation condition. Change as voltage increases indicates moisture or corona in windings
Excitation Current	Tests for short-circuited turns in the windings and for core problems.	Based on initial excitation current test values for transformer. Subsequent test values should be within +/- 5% of initial values
Turns Ratio Test	Measures the transformer winding ratio. If ratio is not the same as calculated, some windings may have short-circuited turns.	Deviation of less than 0.5% from calculated value or adjacent coils

Chapter 7

In general, these tests verify the insulation condition of the transformer windings. Table 7-1 identifies the specific purpose of each test and the acceptable test values.

Although the "megger" test is the most common method of checking insulation integrity, it is a DC test and is therefore not particularly well-suited for testing transformers. This is because the transformer equivalent curcuit is a distributed inductance/capacitance circuit which complicates the DC voltage/current distribution and often leads to an inaccurate reading of insulation resistance. DC testing of a transformer also creates an electrostatic charge on the insulating oil. This charge is not easily dissipated, and can contribute to overstressing the insulation when the transformer is returned to normal service. The power factor and power factor tip-up tests are the best tests for detecting low insulation resistance in the transformer, for two reasons. First, as AC tests, they have a voltage/current distribution very similar to the operating condition of the transformer. Second, they are performed in several directions—low-voltage winding to ground, low-voltage winding to high-voltage winding, and high-voltage winding to ground—and are thus more comprehensive than the "megger" test.

A functional test of the Buchholz relay for conservator-type tranformers also should be performed to confirm that the alarm and trip contacts operate sequentially. All other alarm and trip relays should also be tested regularly to confirm their proper operation.

Infrared Scan. Infrared testing is a standard maintenance procedure that utilizes a heat-sensitive video camera to identify hot spots. An infrared scan taken while the transformer is operating at rated kVA output will identify several developing problems, including:

- Hot high voltage/low voltage termination points caused by corrosion;
- Hot spots on the bushings caused by high leakage currents;
- Hot fan or oil pump motors;
- Plugged radiators;
- Plugged cooling connections;
- Low oil level; and
- Tank hot spots.

The infrared scan of the cooling system is especially important for detecting tube blockage in double-tube oil-to-water heat exchangers used on forced oil/water (FOW) type transformers. This type of transformer uses water to cool the transformer oil and is normally equipped with two cooling pipes enclosing the oil pathway. The presence of an extra pipe enclosing the oil tube makes detecting blockage by other means very difficult.

Oil Testing. Oil circulating throughout the transformer serves as the transformer's cooling medium. Testing the oil for contaminants or dissolved gases provides a means for detecting flaws or hot spots in any part of the transformer. Most

TABLE 7-2: ***ASTM Test Values for Transformer Oil***

Property	Specification Value	Typical Value
Color (with color chart)	0.5 maximum	0.5
Dielectric breakdown at 60 Hz, 0.100 inch gap	30 kV minimum	35 kV
Neutralization number	0.03 mg. maximum	0.01 mg
Viscosity, max centistokes at 0° Celsius	76 seconds maximum	70 seconds
Moisture content	35 ppm maximum	20 ppm
Specific gravity at 60° Celsius	0.910 maximum	0.890
Inorganic chlorides or sulfates	None	None
Interfacial tension	40 dynes/cm minimum	45 dynes/cm
Power factor at 60 Hz and 23 degrees Celsius	0.05% maximum	0.01%

Adapted from ASTM D-3487 [2].

utilities perform oil tests annually, but some prefer more frequent testing to help detect problems early in their development. When extracting oil from the transformer for testing, extreme care must be taken to prevent contamination during the extraction.

The American Society of Testing Materials (ASTM) has outlined normal tests and test values of the oil properties [2]. Normal oil tests include dielectric breakdowns, oil moisture content, interfacial tension, and others (Table 7-2).

An additional oil test is the dissolved gas analysis (DGA). This test, which became prevalent in the mid-1970s, is an analysis of the gases present in the oil. Minor faults within the transformer can cause low-level arcing or local overheating, which will result in the formation of combustible gases such as ethylene, hydrogen, or acetylene, or other gases such as carbon monoxide or carbon dioxide. These gases dissolve in the insulating oil. Therefore, the presence of a higher-than-normal concentration of these dissolved gases in the oil indicates a problem in the transformer. Although the Institute of Electrical and Electronics Engineers (IEEE) has defined "normal" quantities, the actual normal level of gas concentration can vary considerably from tranformer to transformer.[3] Continuing discussions within the IEEE and the International Electrotechnical Commission (IEC) concern the level of carbon monoxide or dioxide that may indicate a transformer problem. Therefore, baseline "normal" values must be established for each transformer before testing for deviations.

Chapter 7

Changes in Oil Properties. A careful comparison of the oil property test values and dissolved gas concentrations to previous test results is essential. If the test results differ significantly from previous results, a retest by a second testing firm is warranted to check the original results' validity. If the second tests confirm the presence of water or a contaminant, the oil must be reconditioned by vacuum treatment. Reconditioning is normally performed by an outside contractor because of the specialized facilities required.

Insulator Maintenance Testing

Insulator maintenance testing consists of separate tests that are applied only to the transformer's high-voltage or low-voltage terminal insulators.

The materials used in bushings are selected to minimize the leakage current, provide a long life while operating in contaminated areas, and meet the necessary mechanical torsion forces. Main power transformer bushings are made of wet-process porcelain, which is actually a blend of different types of clay, feldspar, and quartz. "Wet process" describes the method and amount of water used in the process of manufacturing the porcelain. Bushings rated above 69 kV have a large physical size, which also helps to limit the amount of leakage current during rain-storms or snowstorms, or when the bushing has become contaminated by dust, dirt, salt spray, or other airborne materials.

Power Factor Test

In the power factor (PF) test the high-voltage lead of the PF test equipment is connected to the bushing terminal of the high-voltage bushing and the low-voltage lead is connected to the bushing flange, which must be isolated from the tank, or to the bushing capacitance tap (if furnished) at the bottom of the bush-

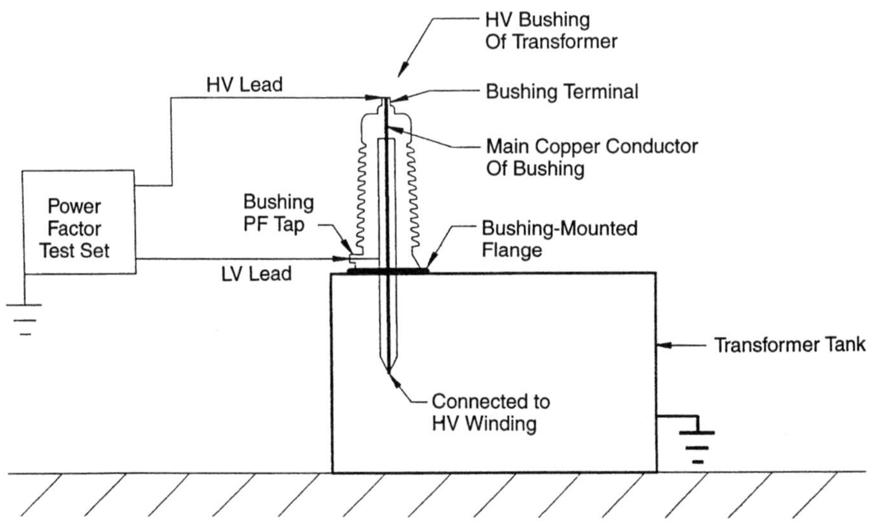

Figure 7-1: ***Ungrounded Specimen (Power Factor) Test.*** *This test is used to measure the performance of the transformer bushing insulation. High voltage is applied to the high-voltage bushing terminal and the leakage current is measured at the bushing capacitance or PF tap.*

ing. The high-voltage lead of the PF test set carries only the current of the bushing under test, and all other high-voltage bushings and the tank are earthed. In this way the leakage current through each bushing may be measured separately (Figure 7-1). An increase in the leakage current over previous tests indicates that the bushing has been damaged or contaminated. However, the test is sensitive to atmospheric conditions such as temperature and humidity, which will affect the normal leakage current in any bushing.

This test is also called the "ungrounded specimen test" (UST) for a transformer bushing because the bushing remains connected to the transformer winding during the test. The low-voltage lead of the PF test set is not connected directly to the ground, as it would be in a true PF test. Instead, it is connected to the bushing capacitance tap or the flange at the bottom of the bushing, both of which are insulated from the tank ground. Generally accepted PF test values range from 100 milliwatts at 10 kV to 6 milliwatts at 2.5 kV.[4]

Hot Collar Test

The hot collar test is usually conducted as a supplement to the ungrounded specimen test on GSU bushings. The hot collar test is used to identify localized defects in the bushing porcelain. The PF test set is used again, but in this test the HV lead is connected to a metal collar fastened around the top porcelain skirt of the bushing, or at various heights on the bushing (Figure 7-2). This test will measure different leakage currents through the porcelain, depending on where the hot

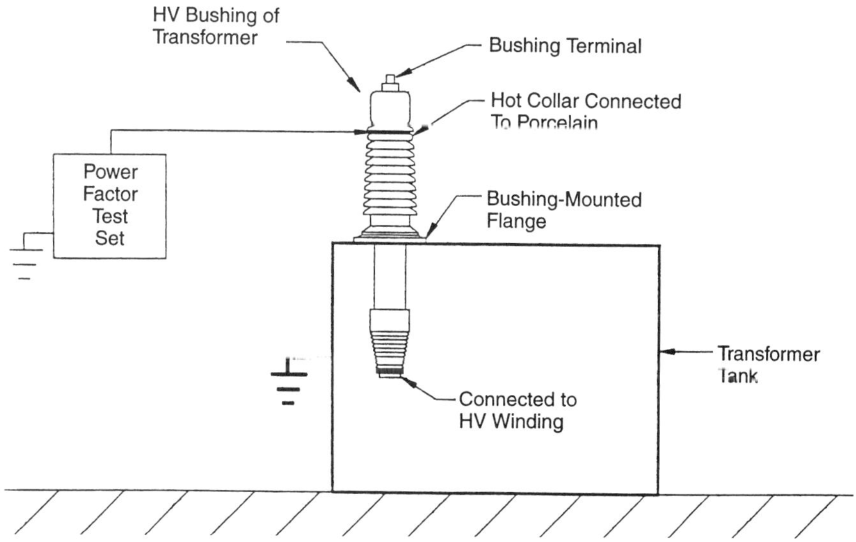

FIGURE 7-2: *Hot Collar Test.* *The hot collar test is used to detect localized defects in the bushing. Current is applied to a collar connected around the bushing at various positions and the leakage currents at each position are compared.*

Chapter 7

collar is situated on the bushing. Differences in leakage current from one area of the porcelain to another area can indicate a low oil level within the bushing or a local defect in the porcelain. This test is especially useful in bushings that lack oil sight glasses or level gauges.

Oil Tests

Before the mid-1980s, plant owners typically monitored only the oil level of the bushing. More recently, however, accepted maintenance practices have come to include a chemical analysis of the bushing oil once every five years. These analyses can help to reveal defects—for example, internal hot spots—before a failure occurs. The analysis is similar to the laboratory dissolved gas analysis performed on the oil in the main transformer tank. Bushing oil samples are taken through pilot valves at the base of the bushing. Many older bushings, made before the mid-1980s, lack these valves.

Bushing Infrared Tests

An infrared scan of the bushings when the transformer is operating at rated MVA should identify any hot spots in the bushing or the bushing terminations. Hot spots on the bushing may indicate a low oil level or mechanical failure of the porcelain. Mechanical failures of the porcelain allow excess leakage current and result in higher-than-normal dielectric losses. (These losses would also be evident from the power factor test values obtained in the ungrounded specimen test or hot collar test.)

Infrared hot spots on the terminations usually indicate that the bolt connections are loose, or that corrosion has occurred somewhere in the termination. Infrared scans are especially important for the high-current bushings (usually with a current greater than 5 kA) where a great number of terminal flanges and connecting bolts are used to facilitate termination cooling. In some cases, depending on the arrangement of the winding conductors in the tank and the low-voltage terminal flanges, excess stray fluxes may also cause localized overheating on the low-voltage termination.

ANSI Standard C76.1 restricts the hot spot temperature rise of the current-carrying parts of the bushing to 65 degrees Celsius over a maximum ambient air temperature of 40 degrees Celsius, or 105 degrees Celsius overall temperature.[5]

Tests and Inspections of Other Components

Conservator

If the transformer has a conservator, moisture tends to accumulate over time at the bottom of the conservator tank. This is caused by condensation on the inside wall of the tank. This moisture should be removed periodically by opening the drain valve provided for this purpose.

Most conservators manufactured after the early 1980s have a plastic bag inside the conservator tank which is connected to the dehydrating silica breather. This bag expands and contracts as the temperature of the oil increases and decreases. The bag eliminates any contact between the atmosphere and the transformer, except through the silica breather, and prevents moisture or other atmospheric agents from contaminating the oil. However, water that has condensed between the tank wall and the plastic bag can still enter the bag through osmosis, contaminating the oil. This problem can be averted by draining the condensed moisture and checking the functioning of the drain valve on an annual basis.

On-Load Tapchangers

On-load tapchangers (OLTCs) are usually maintenance-intensive equipment. The OLTCs are used to vary the transformer's low voltage input, and thus the HV output, while the GSU is supplying load. OLTCs effectively change the GSU's winding ratio. This requires a separate equipment assembly that houses a motor-driven mechanism, reversing and arcing switches, series or booster transformers, and a number of "make-before-break" contacts. This assembly usually holds its own oil supply separate from the main tank oil supply. A good maintenance program should confirm the electrical integrity of each essential component, verify that the number of OLTC operations has not exceeded the life curves of the make-before-break contacts, and include oil testing for the OLTC oil supply.

References

[1] "Guide for Loading Mineral-Oil Immersed Power Transformers up to and Including 100 MVA," ANSI/IEEE C57.92, American National Standards Institute(ANSI)/Institute of Electrical and Electronics Engineers (IEEE), 1981.

[2] "Standard Specifications for Mineral Insulation Oil in Electrical Apparatus," American Society of Testing and Materials (ASTM) Standard D-3487, undated.

[3] "IEEE Guide for the Detection and Determination of Generated Gasses in Oil Immersed Transformers," ANSI/IEEE C57.104, ANSI/IEEE, 1991.

[4] Meyers, S.D., Kelly, J.J., and Parrish, R.H., "A Guide to Transformer Maintenance," Main Transformer Institute, Division S.D. Meyers Inc., Akron, OH, 1981.

[5] "IEEE General Requirements and Test Procedures for Outdoor Apparatus Bushings," ANSI/IEEE C76.1, ANSI/IEEE, 1976.

Bibliography

"Guide for Acceptance and Maintenance of Insulating Oil in Equipment," ANSI/IEEE C57.106, ANSI/IEEE, 1991.

Chapter 8
International Test Standards for Transformers

A hydropower project owner purchasing transformers can choose from high-quality products produced in North America, South America, Europe, or Asia. Although the equipment produced by major manufacturers has a high degree of consistency, the performance that can be expected by a project owner or operator can vary in important ways according to the test standards used in manufacture and procurement. Two sets of standards are preeminent internationally: the American National Standards Institute (ANSI) standards and the International Electrotechnical Commission (IEC) code. There are many similarities between the two codes; however, their differences can have a substantial effect on both the buyer and seller of large power transformers.

The applicable ANSI codes are ANSI Standards C57.12.00 and C57.12.90.[1,2] The IEC code addressing standards for power transformers is IEC Standard 76.[3] Some of the differences between the two codes that can have significant implications for the equipment's performance are detailed in this chapter.

Standards for Transformer Efficiency (Losses)

The ANSI standards allow zero error tolerances (averaged over all units) during measurement of load and no-load losses in transformers, if two or more transformers are ordered. If only one transformer is being purchased, the no-load loss error tolerance can be 10 percent and the load loss error tolerance can be 6 percent.[1] The IEC standards state that a 10 percent error tolerance is acceptable during measurement of the no-load and load losses for each transformer, regardless of the number of transformers ordered.[3] That seemingly small difference can have important implications for the performance of the equipment. It can also significantly affect the cost to both manufacturer and owner in the typical case where a purchase agreement provides for monetary penalties if the transformers do not meet specification requirements.

Consider a hydro project owner who orders three GSU 400-MVA transformers under a contract that specified guaranteed load losses not to exceed 3,300 kW and a penalty of $3,000 per kW for failure to meet the specifications. Factory tests show the three transformers with load losses of 3,350 kW, 3,280 kW, and 3,330 kW, respectively. If the equipment specifications relied on the IEC standard, the load losses of each of the transformers would be within the 10 percent tolerance and no penalties would be assessed. However, if the contract specified ANSI standards, the equipment would be assessed penalties. Using ANSI-prescribed methods, the average load loss of 3,320 kW for the three transformers would exceed the contract allowance of 3,300 kW per transformer by 20 kW, and result in a penalty of $60,000 for the three transformers.

Many project owners and consultants dictate specifications that allow zero tolerance on no-load and load losses, regardless of the number of transformers ordered. In that case, transformer losses such as those noted above would have generated penalties of $240,000 [(50 kW + 30 kW) x ($3,000/kW)].

Lightning Impulse Test Standards

The ANSI and IEC test codes specify the same standard wave shape for the lightning impulse test: 1.2/50, or 1.2 microseconds to reach the peak voltage value in kV and 50 microseconds to reach 50 percent of that same value. However, the ANSI and IEC codes suggest different sequences for performing the various forms of lightning impulse tests (full wave, reduced full wave, chopped wave, and reduced chopped wave).

ANSI Test Sequence

ANSI's normal sequence of lightning impulses applied to each high-voltage terminal of a large power transformer is shown in the first column of Table 8-1. This sequence assumes that no front-of-wave (FOW) impulses are required as direct lightning protection in the form of shield wires or air terminals will be provided at the transformer location. If the transformer is provided with a high-voltage tap-changer that uses non-linear resistors, the high-voltage terminal reduced full wave (RFW) and reduced chopped wave (RCW) voltage waveforms will not match the full wave (FW) and chopped wave (CW) waveforms because of the non-linearity of the resistor as the voltage increases. (A variation of voltage waveform usually indicates an internal transformer fault.) Consequently, the ANSI code states that the lightning impulse test sequence should follow the sequence shown in the second column of Table 8-1, if resistors or other non-linear protective devices are connected across portions of the windings.[2] The ANSI code also stipulates that the first RFW and RCW should be matched against the last RFW and RCW to ensure no waveform deviation has occurred that would indicate an internal fault. Because the second sequence takes longer to perform than the first, the project

TABLE 8-1: ***Lightning Impulse Test Sequence under ANSI Code***

Normal Sequence	Sequence with Non-Linear Resistors
RFW	RFW (62.5 percent)
RCW (optional)	RFW (80 percent)
CW	FW
CW	RCW (62.5 percent)
FW	CW
	CW
	RFW (80 percent)
	RCW (62.5 percent)

Terminology: FW, full wave; RFW, reduced full wave; CW, chopped wave; RCW, reduced chopped wave.

Source: ANSI/IEEE C57.12.90 [2]

TABLE 8-2: ***Lightning Impulse Test Sequence under IEC Code***

Normal Sequence	Sequence with Non-Linear Resistors
RFW (62.5 percent) (optional)	RFW (62.5 percent) (optional)
	RFW (80 percent)
FW	FW
RCW (62.5 percent)	RCW (62.5 percent)
CW	CW
CW	CW
FW	FW
FW	FW
	RFW (80 percent) (optional)
	RFW (62.5 percent) (optional)

Terminology: FW, full wave; RFW, reduced full wave; CW, chopped wave; RCW, reduced chopped wave.

Source: IEC 76-3 [3]

owner/engineer should be prepared to schedule additional time for the test if non-linear resistors are involved.

IEC Test Sequence

The IEC code, like the ANSI code, requires different test sequences for the normal and non-linear resistor arrangements, with more tests being required for equipment with non-linear resistors. Both sequences differ slightly from the ANSI sequences, as shown in Table 8-2. Although reduced wave tests at the end of the sequence are not mandatory under the IEC code, most manufacturers add a minimum of two RFWs at the end of the test for comparison purposes.

Tolerance Allowed for Lightning Impulse Tests

Both ANSI and IEC standards require lightning impulse waveforms to have certain waveshapes. For example, the FW must have a waveshape of 1.2/50 microseconds. Moreover, oscillations on the impulse voltage waveform must not exceed 5 to 10 percent in the vicinity of the crest or 25 percent at the end of the impulse. However, the standards recognize that the generation of high-voltage impulse waveshapes is not an exact science. As a result, both ANSI and IEC allow rather large tolerances for the impulse waveshapes, as shown in Table 8-3. The tolerances should be considered maxima and not be exceeded in most cases. Both ANSI and IEC have general clauses stating that the impulse waveforms may not be attainable depending on the electrical equipment tested.[2,3] The ability to produce precise waveforms also depends on the quality of the manufacturers' test equipment. When equipment characteristics preclude meeting the IEC- or ANSI-specified waveform tolerances, wider tolerances may be agreed on by the manufacturer and the purchaser.

TABLE 8-3: ***Tolerances for Impulse Waveshapes under IEC and ANSI Codes***

Shape Variable	Switching Impulse	Lightning Impulse
Crest Value (kV)	+/– 3%	+/– 3%
Front Time (seconds)	+/– 20%	+/– 30%
Time to Half Value (seconds)	+/– 60%	+/– 20%

Source: ANSI 57.12.90[2]; IEC 76-3 [3]

Polarity for Lightning Impulse Waveforms

ANSI normally requires that lightning impulses be a negative polarity "to reduce the risk of erratic external flashover." In addition, the IEC code also states that a negative polarity lightning impulse waveform is preferred for the FW or CW lightning impulse waveforms. In fact, almost all countries require that the FW and CW waves utilized for lightning impulses be of negative polarity. This reflects two factors: positive impulses flash over at significantly lower voltage levels than negative impulses; and 90 percent of lightning strikes are cloud negative to earth and initiated from the cloud formation overhead, thereby inducing a negative polarity surge in the electrical equipment nearby. Consequently, project owners and engineers typically select a negative-polarity impulse waveform for an FW/CW lightning impulse.

Standards for Switching Surge Impulse Tests

Nominal switching surge impulses have a standard waveshape of 250/2,500 microseconds (250 microseconds to the peak voltage value in kV and 2,500 microseconds to 50 percent of that value).[4] For the switching surge impulse related to transformers, the ANSI code states: "The actual time to crest shall be greater than 100 microseconds and the time to the first voltage zero on the tail of the wave shall be at least 1,000 microseconds."[2] The decreased time period for transformers is provided because the standard switching surge wave-shape applies to equipment that does not have a magnetic circuit that can saturate, such as a transformer.

Under ANSI standards, positive, negative, or both kinds of polarity waves may be used for switching surge impulses in large power transformers. The IEC rules provide that only negative polarity impulse waveforms are to be used for switching surge impulses. Most engineers will select a negative polarity impulse for the switching surge test as well as the lightning impulse test.

References

[1] "General Requirements for Liquid-Immersed Distribution, Power, and Regulating Transformers," ANSI/IEEE Standard C57.12.00, American National Standards Institute(ANSI)/Institute of Electrical and Electronics Engineers (IEEE), 1987.

[2] "Test Code for Liquid-Immersed Distribution, Power, and Regulating Transformers and Guide for Short-Circuiting Testing of Distribution and Power Transformers," ANSI/IEEE Standard C57.12.90, ANSI/IEEE, 1987.

[3] "Power Transformers, Insulation Levels and Dielectric Tests," IEC Standard 76, International Electrotechnical Commission, 1988.

Bibliography

Creed, Frank C., *The Generation and Measurement of High Voltage Impulses*, ISBN 0-944954-00-6, Center Book Publishers, 1989.

Kuffel, E., and W.S. Zaengl, *High Voltage Engineering Fundamentals*, ISBN 0-08-024213-8, Pergamon Press, 1982.

Chapter 9
Dry Transformers at Hydroelectric Plants

Auxiliary power systems at hydroelectric plants have historically used liquid-immersed transformers. However, a growing number of plant owners and engineers now design plants with dry-type transformers in order to reduce fire hazards in auxiliary power systems. The decision to replace a liquid-immersed transformer with a dry-type transformer in the initial design has a price: higher unit cost and a need for specific expertise in dry-type transformers in detailing transformer specifications. However, choosing dry-type transformers offers significant rewards in personnel and equipment safety as a result of the decreased fire hazards associated with these transformers.

Alternative Liquid-Immersed Transformers

Voltages on hydroelectric plant auxiliary power systems can range from a 20 kV medium-voltage system to the normal 480-208/120 V low-voltage system. Where the transformer is liquid-immersed, it may use a less-flammable liquid than oil to decrease the fire hazard. Less-flammable liquids in use have firepoints of 300 degrees Celsius or higher. Among them, silicone fluids have a firepoint of 344 degrees Celsius and excellent electrical insulating capabilities. High temperature hydrocarbons (HTH) have a firepoint of 312 degrees Celsius and good insulating properties. Both types of chemicals have adequate heat-transfer capabilities. (However, oil's heat-transfer capabilities are better.)

The National Electrical Code (NEC) has less stringent construction requirements for installation of transformers with less-flammable liquids than for oil-filled units. The NEC says the transformer must be installed in a three-hour fire-withstand vault in a liquid-containment area with an automatic fire-extinguishing system.[1] Transformers using less-flammable cooling liquids cost approximately 40 percent more than equivalent oil-filled transformers, which largely reflects the bigger physical size required due to the reduced heat-transfer capabilities of the cooling fluid. Some of this cost, however, is offset by the less-stringent installation requirements. While less flammable than oil-immersed transformers, these liquid-filled transformers still present some hazard.[2,3]

Dry Transformers

For a hydro plant owner or engineer seeking alternatives to liquid-filled transformers, there are two main kinds of dry-type transformers to consider: vacuum-pressure impregnated (VPI) and cast coil epoxy resin. VPI transformers normally are manufactured in sizes up to 6,000 kVA with primary voltage ratings up to 20 kV. Cast coil transformers can be made in primary voltage ratings up to 34.5 kV. Although the same transformer theory applies to dry transformers as to liquid-filled units, the cooling medium, air, requires different temperature rise ratings, dielectric tests, and construction techniques to ensure reliability. Consequently, the factory and field tests for dry units are established by a separate set of American National Standards Institute (ANSI)/Institute of Electrical and Electronics Engineers (IEEE) standards.[4,5]

FIGURE 9-1: ***Dry Transformers in the Powerhouse.*** *Dry-type transformers offer significant fire-safety benefits when used in hydroelectric plant auxiliary power systems. The transformers may be installed without constructing a vault or providing the liquid-containment and automatic fire-extinguishing systems required for most liquid-cooled transformers.*

Dry transformers have been in use for more than 50 years. VPI transformers have been manufactured for some 40 years. Cast coil transformers, although made for more than 20 years, have only become accepted in the U.S. and Canada since the late 1980s.

Construction and Fabrication

VPI Transformers. The process of impregnating a VPI transformer with epoxy resins begins by preheating the entire unit to release air and moisture under vacuum. A vacuum is established, and resin is added to the vacuum chamber to completely immerse the transformer. Pressurized air is applied at approximately 80-100 pounds per square inch (psi) to force the resin into the transformer coils. The pressure is removed, the excess epoxy is drained, and the transformer baked to affix the remaining resin.

The VPI process deposits a thin, uniform coating of epoxy resin over the top of the transformer windings and impregnates the epoxy into the windings to provide moisture protection and some protection against contaminants. The times that elapse for each of the four steps are determined by the individual manufacturer based on the transformer geometry/construction and type of epoxy resin used. As a result, dry transformers should be purchased from a reliable manufacturer with proven products. [6,7]

Cast Coil Transformers. During fabrication, the coils of a cast coil transformer are locked into a permanent mold using a vacuum-molding process. In most cases, both the high-voltage and low-voltage windings are placed in a cast coil. However, placing only the high-voltage winding in a cast coil and using VPI on the low-voltage winding also has become an accepted fabrication method. The rigid mold gives the transformer superior short circuit strength, impulse ratings equivalent to liquid-immersed transformers, and excellent long-term resistance to moisture and atmospheric contaminants.

Choosing Between Cast Coil and VPI Transformers

Cast coil transformers have several important advantages over VPI units.

Because the high-voltage or high-voltage/low voltage coils are encased completely in an epoxy mold, cast coil transformers can operate in much harsher environmental conditions than VPI units. They often are used in cement or chemical plants where dirt, dust, and corrosive atmospheres are prevalent.

The basic impulse insulation levels (BIL) of a cast coil transformer can equal those of a liquid-filled transformer, while a VPI transformer's BIL is usually less. A higher BIL for a nominal voltage means the insulation should last longer. Although ANSI/IEEE standards do not distinguish between the BIL required for VPI or cast coil transformers, a number of the cast coil manufacturers claim the BIL for the high voltage windings of their products are equal to those of liquid-filled transformers.

Cast coil transformers have superior fire resistance. Such transformers have been tested extensively, and the epoxy resin materials they use are classified as "self extinguishing." Moreover, combustion byproducts have been analyzed in detail to ensure that no significant quantities of any toxic materials are released during combustion, thereby ensuring that another Askarel episode will not occur. (Askarel was a nonflammable liquid substituted for oil as a cooling medium in transformers in the 1970s. It was promoted as reducing the fire hazards associated with liquid filled transformers located indoors. Unfortunately, about 70 percent of Askarel was polychlorinated biphenyls, or PCBs. PCBs pose a serious environmental hazard and their use is now prohibited. When a fire occurred in an Askarel transformer due to an internal short circuit, the PCBs were carried into the air by soot and ash. Askarel transformer fires rendered several buildings uninhabitable for years due to the spread of PCBs throughout the building.) Some epoxy resins create great amounts of soot and ash when burned, creating a different type of environmental clean-up issue. Many of the burn products are also toxic (for example, benzoic acid and phenols) but are easier to clean up and dispose.

A cast coil transformer usually requires less maintenance than a VPI transformer, because the windings are mechanically protected by the mold and because of the higher BIL.

Cast coil units also tend to have lower total power losses due to lower I^2R losses. High-voltage windings in such transformers are encased in the mold and do not have air ducts. (Low-voltage windings do have air ducts.) Consequently, the high-voltage windings are kept cool by reducing the I^2R losses with larger copper conductors. The lower power losses result in increased transformer efficiency.

Cost is the main disadvantage of cast coil transformers. Cast coil transformers usually cost $30 to $40 per kVA compared with $20 per kVA for VPI transformers. This should be compared to a cost of $12 per kVA for oil-filled units. Cost will vary slightly, of course, depending on the kVA size.

Temperature Rise in Dry Transformers

The average winding temperature rise for a VPI dry transformer is 150 degrees Celsius, based on a maximum ambient temperature of 40 degrees as required by ANSI/IEEE standards.[8] This rating provides for a 30-degree Celsius maximum temperature gradient in the windings from bottom to top. The gradient commonly is called the "hot-spot temperature." Therefore, a transformer with a 150-degree Celsius temperature rise operating in a 40-degree Celsius ambient temperature with a 30-degree Celsius hot-spot would be able to heat up another 80 degrees and still be within its temperature-rise rating (40 + 30 + 80 = 150).

The temperature-rise rating for dry transformers can be contrasted with liquid-filled transformer temperature rises of 55 and 65 degrees Celsius. Oil and less-flammable liquid cooling mediums clearly can conduct the heat away from a transformer core and windings more efficiently than the air in dry transformers.

VPI dry transformers also can be specified with temperature rises of 80 or 115 degrees Celsius. These temperature rises are specified for low-loss, high-efficiency transformers with longer life expectancies than the normal 150-degree Celsius units. Naturally, a lower temperature rise,while maintaining the maximum ambient and hot-spot ratings set out by ANSI/IEEE, requires more copper in the windings to limit the I^2R losses and a larger enclosure to aid air ventilation. These requirements result in a substantial extra cost. In return for the extra cost, the owner of a 115-degree temperature-rise transformer gets a theoretical life expectancy of about ten times that of a 150-degree transformer and the ability to operate the unit continuously at 115 percent of its nominal rating. An 80-degree Celsius temperature-rise transformer has an even greater life expectancy than the 115-degree transformer and the ability to operate continuously at 130 percent of its nominal rating.

Cast coil transformers are manufactured with a temperature rise of only 80 degrees Celsius. This lower temperature rise, when compared with the normal 150-degree temperature rise of a VPI unit, indicates higher efficiencies and greater overload capabilities. The difference also is reflected in the higher cost associated with cast coil transformers.

Measuring the Performance of Dry Transformers

Construction and test procedures for VPI and cast coil transformers are significantly different than those for liquid-filled units. Transformer users should become familiar with the appropriate ANSI/IEEE standards to ensure a good technical specification with adequate factory testing is prepared. The engineer or owner also should witness factory tests to ensure an acceptable product is shipped from the factory.

Temperature Rise Test

Dry-type transformers usually are constructed in smaller kVA sizes than liquid-filled ones. There are a greater number of manufacturers in the dry-type market, making the market competitive and the manufacturers cost-conscious. Because it uses less energy and requires less extensive testing facilities, the "separate load loss and excitation test" referenced in ANSI/IEEE Standard C57.12.91 is the most commonly used temperature-rise test for dry transformers.[4]

In the "load-loss" portion of that test, enough voltage is applied to the high-voltage terminals to produce the rated current with the low-voltage terminals short-circuited. The transformer temperature has to be stabilized to a 2-degree-Celsius band for a three-hour period before the temperature rise of the windings is determined by the resistance method. In the "excitation" part of the test, the high-voltage terminals are open-circuited, and rated voltage is applied to the low-voltage terminals. The transformer temperature again must be stabilized to a 2-degree-Celsius band for a three hour period. After the core is de-energized, periodic temperature readings are taken and used to calculate the core temperature just before de-energization. The total temperature rise of the transformer then is calculated using the following formula:

$$T_t = T_c[1 + (T_E/T_c)^{1.25}]^{0.80}$$

where:

T_t is the total winding temperature rise with full load current in the winding and normal excitation on the core;

T_c is the high-voltage or low-voltage winding rise measured immediately following the load-loss heat run with full load current flowing in one winding and the other winding short-circuited;

T_E is the high-voltage or low-voltage winding rise measured immediately following the heat run with normal excitation on the core.

Due to the relatively recent introduction of cast coil transformers into the U.S. and Canada, different hot spot temperatures currently are allowed in the International Electrotechnical Committee (IEC) standards than in the ANSI/IEEE standards. Investigations have shown that the ANSI/IEEE standards are more conservative and should be used where possible.[9]

Dielectric Tests

Three "routine" dielectric tests are required for dry transformers: applied voltage, induced voltage, and partial discharge. The voltage levels used for the applied voltage and induced voltage tests are lower than levels used for equivalent liquid-filled transformers.[5] The standards do not require a cast coil transformer to pass a more severe dielectric test than a VPI transformer, regardless of the manufacturer's claims that cast coil transformers have BIL equivalent to those of liquid-filled transformers. Therefore, an owner or engineer who wants the cast coil transformer to pass the voltage test levels for a liquid-filled transformer must specifically state this in the technical specification.

The partial discharge test is listed as a "routine" test for all dry transformers. However, the ANSI/IEEE standard has footnotes associated with it. One says the test may be performed on the windings of all types of dry transformers but is considered "routine" for transformers above 1.2 kV having solid cast and/or resin-encapsulated windings. The other note provides that a partial discharge test may be performed as a "design" test only, where a specific cast coil dry transformer design has been shown to have longer service life.[5]

A prudent hydro plant engineer or owner will perform a partial discharge test routinely on all dry transformers as a standard quality check. In addition, users should specifically require the test on the high-voltage windings of all cast coil transformers greater than 50 kVA to ensure the integrity of the epoxy resin impregnation. The test is necessary because the encapsulation of the high-voltage winding in epoxy resin mold makes it impossible to see indications of partial discharge within the winding after the mold is applied.

Insulation Power Factor Test

Many engineers and owners also require insulation power factor tests as a standard test on all dry-type power transformers.

This test applies an AC voltage to the transformer windings and measures the leakage current through the insulation to ground. It generally is thought that the test values recorded can be used after installation of the transformer and throughout its life as a maintenance parameter to determine if the insulation has deteriorated significantly from its condition at the factory. Although this description of the insulation power factor test is valid for liquid-filled transformers, standard insulation power factor values have not been established for dry transformers.[4] Operating experience has indicated little or no relationship between the insulation power factor test values in a dry transformer and the ability of the transformer to withstand the dielectric tests prescribed by ANSI/IEEE. Consequently, insulation power factor values are used more as a quality assurance comparative value to monitor consistency of the insulation quality from transformer to transformer.

ANSI/IEEE standards identify the insulation power factor test as a design test not normally performed on dry transformers prior to shipment.[5] This seems appropriate. In fact, the author attended one factory test of a cast coil transformer (19.32 kV/480V, 2,500/3,333 kVA), required by the owner's technical specification, where the leakage current through the transformer winding insulation was so low that the voltmeter/ammeter/wattmeter method prescribed for the test could not be used. In place of this method, a more sensitive, capacitance bridge method was required to record the test values.

References

[1] National Electric Code, Section 450-23, National Fire Protection Association, 1993.

[2] Claiborne, C.C., and H.A. Pearce, "Transformer Fluids," *IEEE Electrical Insulation*, Volume 5, No. 4, August 1989.

[3] Obeck, Tor, "Fire Hazards-Present Codes and Standards," *IEEE Electrical Insulation*, Volume 6, No. 3, May-June 1993.

[4] "IEEE Standard Test Code for Dry-type Distribution and Power Transformers," ANSI/IEEE Standard C57.12.91, American National Standards Institute (ANSI)/Institute of Electrical and Electronics Engineers (IEEE), 1979.

[5] "IEEE Standard General Requirements for Dry-type Distribution and Power Transformers Including Those with Solid Cast and/or Resin-Encapsulated Windings," ANSI/IEEE Standard C57.12.01, ANSI/IEEE, 1979.
[6] Richon, George, "Vacuum Pressure Impregnation and the Resin Manufacturer," *IEEE Electrical Insulation,* Volume 2, No. 4, July 1986.
[7] Harrington, Arnold W., "VPI System Considerations When Impregnating Multilayers of Insulation," *IEEE Electrical Insulation,* Volume 2, No. 2, March 1986.
[8] "IEEE Recommended Practice for Electric Power Distribution for Industrial Plants," Standard 141, IEEE Redbook, 1986.
[9] Pierce, Linden W., "An Investigation of the Temperature Distribution in Cast-Resin Transformer Windings," *IEEE Power Engineering Review,* April 1992.

Chapter 10
Controls and Protective Relaying

From the 1920s until the 1970s, control of a hydro plant's generating units was typically performed from a unit control switchboard (UCS), or, if the plant had multiple units, from a plant control switchboard (PCS). The UCS and PCS switchboards contained iron vane meters, hardwired control switches, and hundreds of auxiliary relays to perform the unit start/stop operations. All the necessary sensors and controls required to operate the unit or units were hardwired to the UCS or PCS, allowing a single operator to control the entire station from one location. Although modern systems still permit control of the entire plant from a single location, the UCS is no longer hardwired to each individual field component. Modern control rooms utilize the far more cost-effective supervisory control and data acquisition (SCADA) systems (including programmable logic controllers (PLCs) and distributed computer control systems (DCSs)) with graphic display screens to implement a vast array of control schemes. The SCADA control scheme also provides flexibility in control, alarming, sequence of events recording, and remote communication that was not possible with the hardwired control systems.

Choosing a Control System

Selecting a Supplier

Control technologies continue to change rapidly. Therefore, one of the major items to investigate before installing a modern control system is the longevity of the equipment supplier and his commitment to supporting his existing hardware/software equipment lines. "Longevity" is the supplier's capability to adapt and modify its own existing equipment to accept new changes when new equipment designs and software protocols are introduced into the marketplace. Therefore, the longevity of the vendor ensures that your purchased computer system remains a flexible and updateable system. The longevity of the equipment supplier is also directly related to the marketshare in percent of total market that the equipment supplier controls.

Types of System

The type of control system installed at a hydro plant depends largely on the amount and complexity of data to be manipulated. PLC systems are significantly less expensive than computer-based SCADA systems and also are easier to program and troubleshoot. However, PLC systems' processing speed and scan rate are low, limiting their ability to handle complex control algorithms and large quantities (1,000 or more points) of input/output (I/O). Computer-based SCADA systems (for example, DCS systems) are extremely versatile in handling complex control algorithms and large quantities of information due to their higher processing speed and math-intensive capabilities.

Control schemes for hydro plants vary considerably depending upon the size and complexity of the plant. For small hydro stations with one unit and minimal auxiliaries, a simple PLC scheme with manual back-up control is normally utilized. A computer-based SCADA system is often installed at a large hydro station with multiple units and complex reservoir regulation requirements. In between the simple control system used for a small single-unit hydro plant and the complex control system used for a large multiple-unit plant, an engineering decision must be made concerning which type of control scheme to implement. This decision should be based upon the complexity of plant operations, response time required, and the degree of reliability and automation desired.

Emergency Shutdown Controls

Although computer and PLC systems improve plant operation by providing greater flexibility in control, alarming, and sequence of events recording, the essential emergency shutdown controls should remain hardwired. This will guarantee that a safe and orderly shutdown of the plant can be accomplished in an emergency situation during which the computer/PLC has failed.

Control Systems at Small Hydro Plants

Small hydro plants are defined by the American National Standards Institute (ANSI) and Institute of Electrical and Electronics Engineers (IEEE) as plants that produce 5 MW of power or less. Control functions of small hydro plants are defined in ANSI/IEEE Standard 1020, "IEEE Guide for Control of Small Hydroelectric Plants."[1] This standard defines individual control sequences for starting, stopping, braking, etc. of the unit. However, the specific hardware and software to be utilized for implementation of these control sequences is not addressed.

Types of Schemes

The two basic control schemes utilized for small and medium hydro stations are (1) a single PLC w ith a manually operated back-up system, and (2) a redundant

TABLE 10-1: ***Advantages and Disadvantages of the Redundant PLC Control Scheme***

Advantages	Disadvantages
•100 percent backup for the central processing unit (CPU). The CPU includes the processor, system memory, and system power supply.	•Cost. The cost of a second PLC exceeds the cost of a manual backup system.
•Continued automatic control of the unit under headwater level or discharge control with one PLC out of service. This ability allows continued maximizing unit revenue when a PLC fails.	•Complexity. Most small hydro plant operators are not technically trained for troubleshooting PLCs. (Some of this complexity is offset by the PLC and I/O card self-diagnostics now available.)
•Uniform spare parts. Only one set of I/O cards needs to be maintained. Items such as spare relays and control switches associated with a hard-wired system are not required.	Failure of both systems simultaneously. Although redundant PLCs do enhance system reliability, they can be prone to simultaneous failure caused by surges. Owners should insist on good surge protection engineering.
	•Software problems. If software is non-standard, software problems will be common to both PLCs.

TABLE 10-2: ***Advantages and Disadvantages of a Single PLC With Manually Operated Backup System***

Advantages	Disadvantages
•Less expensive than a redundant PLC system.	•Headwater level or discharge control (if performed by the PLC) is disabled whenever the PLC is disabled. While utilizing the manually-operated backup system for control, the unit's output is set at the operator's discretion. An operator will usually allow a safety margin of approximately 10 percent in headwater or discharge level to avoid problems such as drawing air into the penstock. As a result, maximum possible revenue for the unit is usually not realized during manual operation.
•Less chance of a common mode failure because the hardwired system is less prone to surge-induced failures and more tolerant of inadequate grounds.	•Nonuniform spare parts. Spare parts would have to be provided for both the PLC system and the manually operated backup system. However, it should be noted that relatively few spare parts would be needed for the manual backup system, due to its simplicity.
•Operator familiarity with troubleshooting hardwired relay systems.	

PLC system. There are various modifications of these two basic schemes, which depend upon the individual plant requirements and owner preference. The single PLC offers the advantages of low cost and simplicity, and is typically backed up by a hardwired system. With a redundant PLC system, backup control and memory are provided by a second PLC. Advantages and disadvantages of the two schemes are summarized in Tables 10-1 and 10-2.

FIGURE 10-1: ***Unit Control Switchboard for Moose River Powerhouse.*** *The Moose River plant is a 13.26-MVA, single-unit plant. The unit controls and indication are located in the cubicle to the left, and protective relays are located in the two center cubicles. The far right cubicle contains the PLC, temperature monitor, and annunciator.*

FIGURE 10-2: ***Unit Control Switchboard for McNary Powerhouse.*** *In this 12.0-MVA, single-unit powerhouse, the unit controls and indication are on the left-hand cubicle and the PLC, temperature monitor, and annunciator are located on the right-hand cubicle. Protective relays are mounted on separate cubicles (not shown), which also contain the generator circuit breaker.*

Chapter 10

Figures 10-1 and 10-2 show two typical UCSs for small hydro plants. In either unit control scheme, all unit protective relays should be independent from the programmable controllers. This independence will allow the protective relays to function even if the PLC fails, ensuring the safety of unit equipment and personnel. For the single PLC scheme with a manually operated back-up system, it is usually best to have an independent resistance temperature detector (RTD) monitor and annunciator panel functionally operative during manual operation of the unit. These additional panels will provide the operator vital information which will facilitate operation of the plant in the manual mode.

Personal Computers

Modern control schemes also utilize personal computers (PCs) in conjunction with PLC control systems. The PCs are utilized with man-machine interface (MMI) software for control display graphics, historical data and trend displays, computerized maintenance management systems (CMMS), and remote communication and control. In addition, the PLC programming software is usually resident on the PC, eliminating the need for a separate programming terminal to implement or change the PLC software coding.

A PC also can be used for graphical displays of plant data, greatly enhancing operational control. Standard Microsoft-based graphical display software packages are available for installation on a standard PC. An example of a typical graphical display software package is INTELLUTION's 32-bit FIX software package for Windows 95 and Windows NT. INTELLUTION is one of the top five graphical display software vendors in the U. S. The FIX software package can be utilized on the PC to create specific powerhouse graphical displays based upon real-time PLC inputs. These displays typically include control displays with select-before-execute logic, informational displays for plant RTD temperatures, or historical trending plots of headwater, tailwater, and flow data.

Modems with both dial-out and dial-in capabilities can be located in either the PC, the PLC, or both to provide off-site access to plant information. These modems may also be utilized to control the plant operation from a remote location. A typical control scheme utilizing PCs and PLCs is shown in Figure 10-3.

Control Systems at Medium and Large Hydro Plants

The complex operating systems at many large hydro plants require a computer-based SCADA system computer in place of a PLC control system. Some elements of large plant operation that create this complexity are multiple units; complex reservoir or run-of-river water level control algorithms; and operation of numerous intake gates, inlet valves, spillway gates, sluice gates, and draft-tube gates in conjunction with a unit start/stop sequence or for flood conditions. The DCS is appropriate for this type of operation because it has the ability to handle large

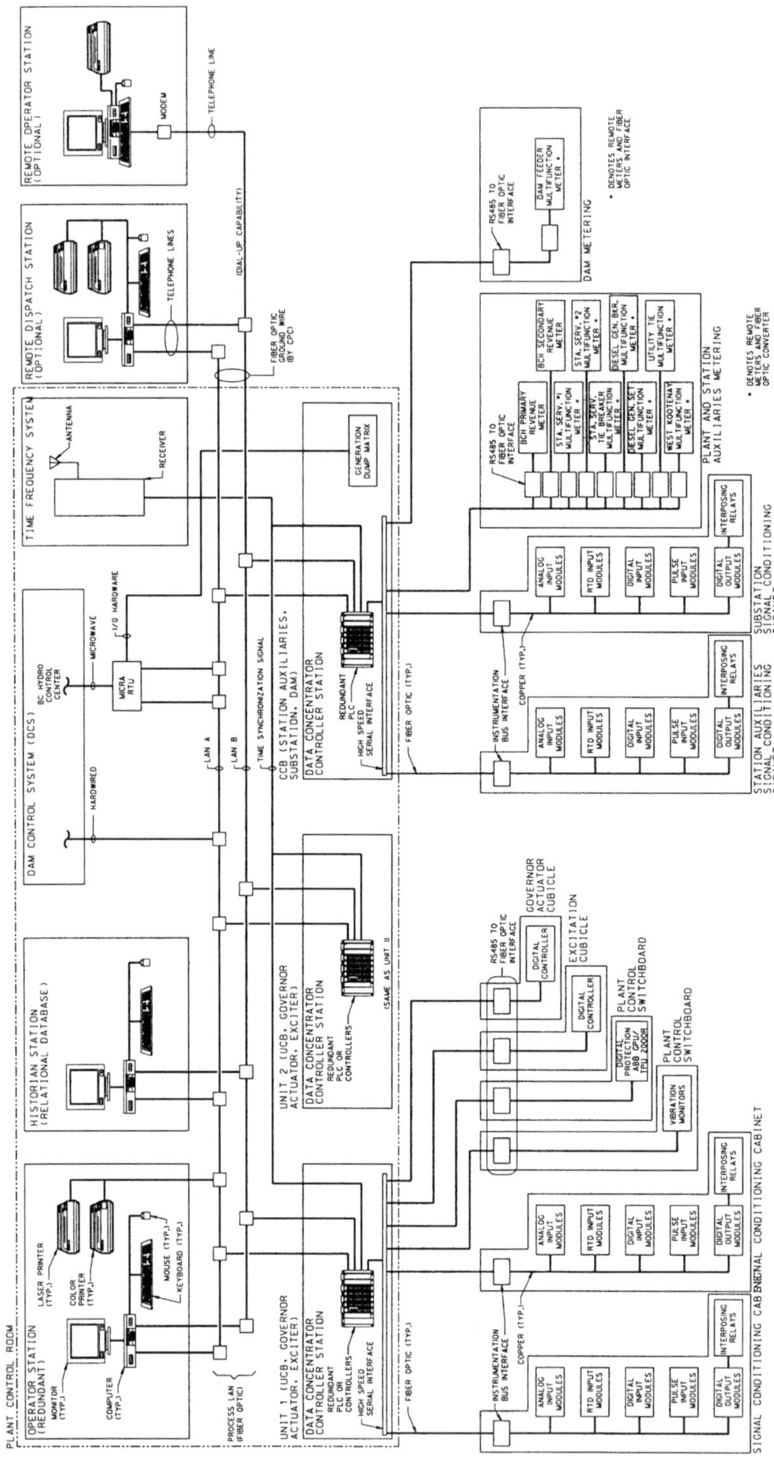

FIGURE 10-3: *PLC/PC Arrangement for the Keenleyside Hydroelectric Power Station.*

quantities of input and output quickly, providing operators with "real-time" information.

The various control hierarchies and standard control sequences for medium and large hydro plants are described in ANSI/IEEE Standard 1010, "IEEE Guide for Control of Hydroelectric Powerplants."[2]

Computer-Based SCADA Systems

Computer-based SCADA systems use microprocessor-based control technology physically distributed throughout the powerhouse. The separate microprocessor units are linked together by a number of digital communication paths to form a completely integrated control system. The distributed system offers several advantages over a centralized mainframe computer type system, including:

- Improved system performance, as the various distributed microprocessors perform dedicated software functions or tasks at the same time (parallel processing);
- Reduced software complexity, with each microprocessor performing its own dedicated tasks;
- Modular elements, making the system easily expandable and simplifying component replacement; and
- Increased system reliability because of the reduced complexity and modular structure, which allows most of the system to function through the failure of one or more components.

Specifying the Response Time

The processing speed of the computer or computers used in a control system determines the overall system response time. The response time of the system is very important, especially in emergency situations. Therefore, when specifying a control system, the plant owner should take care to define the response time speed of the computer system clearly. Some of the items related to the response time of a computer system are:

- Time duration required to update a graphical display from the instant a field contact changes state;
- Time duration from the instant a control is activated at the operator station until the command is implemented at the field device; and
- Overall time duration to process and log an alarm once it is received at the computer.

The owner's specification should define the acceptable time durations for each of the above events in both a normal loading and high-activity loading scenario. Penalties should be included in the specification if the vendor cannot meet the response times defined. The computer system response times should be verified at the factory acceptance test to confirm the system will operate as required by the specifications.

Equipment Architecture and Protocols

Many basic architectures and communication protocols are utilized for computer and PLC systems. This is because historically, the major computer manufacturers (such as IBM, Digital Equipment Corporation, and Xerox) and PLC manufacturers (such as Allen Bradley, General Electric, and Modicon) established proprietary architectures and protocols before any industry standardization was attempted. The result was that until the 1990s, all architectures and protocols were the vendors' proprietary versions and could not easily interface with those of other manufacturers.

Although these various vendor architectures and protocols still exist, the computer industry has been attempting since the late 1980s to define "open" architecture and communication protocol standards to which all computer manufacturers must adhere. Such non-proprietary architectures and protocols would facilitate expansion of computer systems by other manufacturers and allow easier communication between different computers. However, many conflicts and disagreements among various parties have slowed the process, and standardization is not yet complete. Therefore, architecture and communications are two potential problem areas that should be addressed by purchasers of large DCS computer systems (and to a lesser extent PLC systems) for a hydro plant control system.

Open Architecture Systems

"Open architecture systems" and "open systems" are common buzzwords used by many vendors. However, the meaning of "open" is still not well defined. An offer to supply an "open" computer system should define the interface or operating standard for the following:

- Communications;
- Operating system;
- User interface; and
- Databases.

Each of these elements should be investigated for its degree of openness—that is, its capability to be replaced by or communicate with computer system elements provided by other vendors. The vendor should provide sufficient definition of these elements to allow the owner to determine how "open" or non-proprietary the computer system will be when interfacing with other computer systems.

In the early 1990s, the International Organization for Standardization (ISO) developed a model for open architecture and protocol, known as the OSI (open systems interconnections)-ISO model. Designs consistent with the OSI-ISO models will allow upgrades or expansions of the computer system's subsections by manufacturers other than the original seller.

Communication Protocols

No computer system can be defined as "open" if the communication protocol is not open. A communication protocol defines the format and structure of the messages sent from one microprocessor or element in a computer to another microprocessor or element in the computer; and from one computer to another computer.

Any plant control computer-based SCADA system specified for a hydro plant should have an open architecture system with a communication protocol conceptual model consistent with the OSI-ISO reference model. The OSI-ISO reference model is supported by the utility communications architecture (UCA) sponsored by the Electric Power Research Institute (EPRI). The UCA is to be utilized for information exchange by utilities in the U. S. for their process-control and business information systems. The OSI-ISO model does not define the actual protocols to be used but defines the conceptual model for manufacturers to follow.

Many proprietary communication protocols still exist in the current market due to the history of the computer/PLC industry. However, the industry is slowly moving towards communication protocols that are UCA/OSI-ISO compliant.

The OSI-ISO model defines a seven-level conceptual communication system which will permit machines with different established protocols to exchange data without major complications. This is advantageous because many communication protocols already on the market (e.g. Xerox's Ethernet, IBM's SNA, DECNet) are already accepted in North America as *de facto* standards. Two U. S. utility groups, the Western States Coordinating Council (WSCC) and Interutility Data Exchange Consortium (IDEC), also established two different protocol standards prior to the advent of the OSI-ISO model. The protocols developed by the two utility groups are not able to communicate with each other. However, at present, the two utility groups are working together to develop a UCA-compliant protocol profile utilizing OSI.

PROTECTIVE RELAYING

A protective relay is a device that activates an alarm or initiates the interruption of the electric power flow when it senses an abnormal situation that could harm plant personnel or equipment. Protective relay technology has changed significantly in recent years. In the 1980s, the technology progressed from utilizing individual induction-disk relays for each protective function to individual solid-state relays. In the 1990's, microprocessor-based multifunction digital protective relays replaced the individual solid-state relays of the 1980's.

Advantages of Microprocessor-Based Protective Relays

Microprocessor-based digital protective relay technology provides numerous advantages over the individual induction-disk or solid-state technologies. A few of these advantages are listed below.

Self-Monitoring. Microprocessor-based digital relays use "watchdog timers" to monitor their own operating status on a continual basis. Any potential malfunction will be identified and communicated to the DCS/PLC control system. The self-monitoring feature eliminates the possibility of a non-functioning relay in the plant protective relay system. In previous protective relay systems, a non-functional induction disk or solid-state relay would not normally be discovered until the next maintenance test of the protective relay system.

Communication. The digital relay uses a digital communication scheme which allows the relay to communicate directly with the plant control system. The digital relays are provided with serial data ports based on established protocols, which should be compatible with the DCS/PLC communication protocol used at the plant.

Multiple Protective Functions. Digital relays provide multiple protective functions in one relay. In contrast, older relay systems required an individual relay for each protective function. Consequently, multifunction digital relays reduce panel space and wiring costs while providing equivalent protection.

Self-Calibration. Digital relays are provided with a self-calibration routine, which can be initiated by selecting the relay calibration mode in the relay's software programming.

Programmable Setpoints. Previous relay systems required "experienced" protective relay engineers to calculate the setpoints for each individual relay, define the zones of protection for each primary and back-up relay, and perform a coordination analysis to confirm that the operation times of the various protective relays did not conflict. The digital relay uses a DOS- or Windows-based software program which provides tutorials and recommended setpoints for each protective relay function based upon system characteristics. The setpoint tutorials provide assistance in setting the relay. When the basic system data are entered, the setpoint programs will perform the numerous coordination calculations required to provide a coordination analysis. Most software programs also provide graphic display worksheets which show the time-current characteristics of the relays in graph form to simplify coordination between the multiple protective functions that reside in one relay. This ensures that all protective relay functions included in a digital multifunction relay are selectively coordinated.

Event Storage. The digital relays can store eight to 10 selected waveforms on an oscillograph record. The number of cycles that can be stored varies with the manufacturer. This record will show the condition of each of the selected waveforms before and after the protective relay has operated. This additional information is valuable in determining the cause of the protective relay trip.

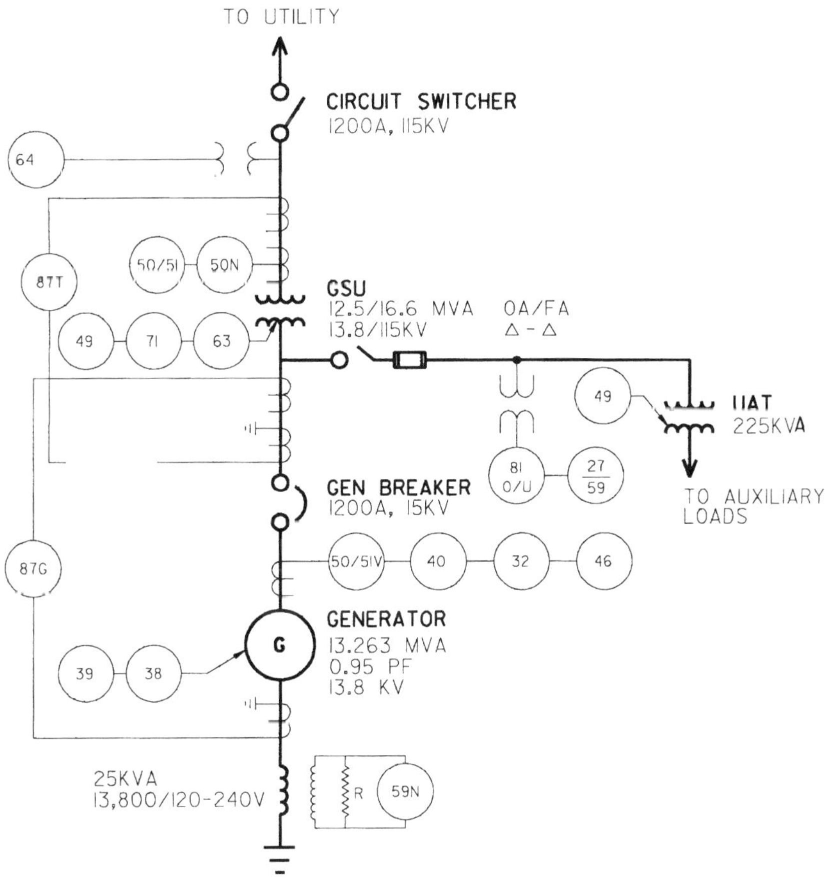

FIGURE 10-4: ***Protective Relay Scheme at the Moose River Project.*** *This schematic shows the protective relay scheme installed in 1987 at a 13.26-MVA plant in New York State.*

Protective Relays at Small Hydro Plants

Typical protective relaying schemes for various types of small hydro plants are identified in IEEE 1020.[1] Some standard protective relay schemes that Harza has utilized for small hydro plants are illustrated in Figures 10-4 and 10-5. The plants illustrated were constructed in 1987 and 1990. Consequently, individual protective relays rather than microprocessor-based digital relays were used for each protective function. However, the protective relaying schemes illustrated exceed the requirements of the IEEE 1020 guide.

Protective Relays at Medium and Large Hydro Plants

ANSI/IEEE Standard 242, "IEEE Recommended Practice for Protection and Coordination of Industrial and Commercial Power Systems" identifies typical

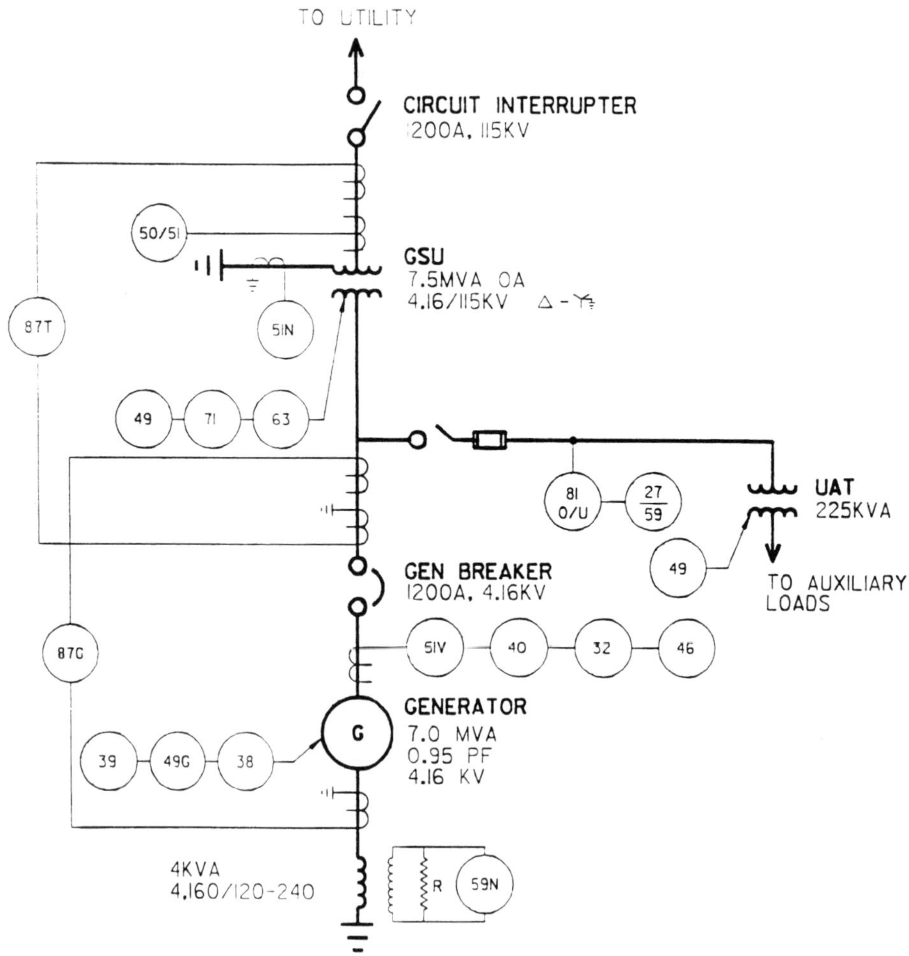

FIGURE 10-5: ***Protective Relay Scheme at the PEC Headworks Project.*** *The protective relay scheme shown here was installed in 1990 at a 7.0-MVA plant in Washington State.*

protective relaying schemes for large electrical equipment items located in medium and large size hydro plants.[3] A typical protective relay scheme for medium and large hydro plants is shown in Figure 10-6.

Equipment Testing

Protective relay systems should be tested regularly to verify the reliability of the relay system. Although the microprocessor-based protective relays' self-monitoring and self-calibrating software has greatly simplified preventive maintenance procedures, a regularly scheduled preventive maintenance test program is still needed to confirm that all relay parameters, setpoints, and communications are

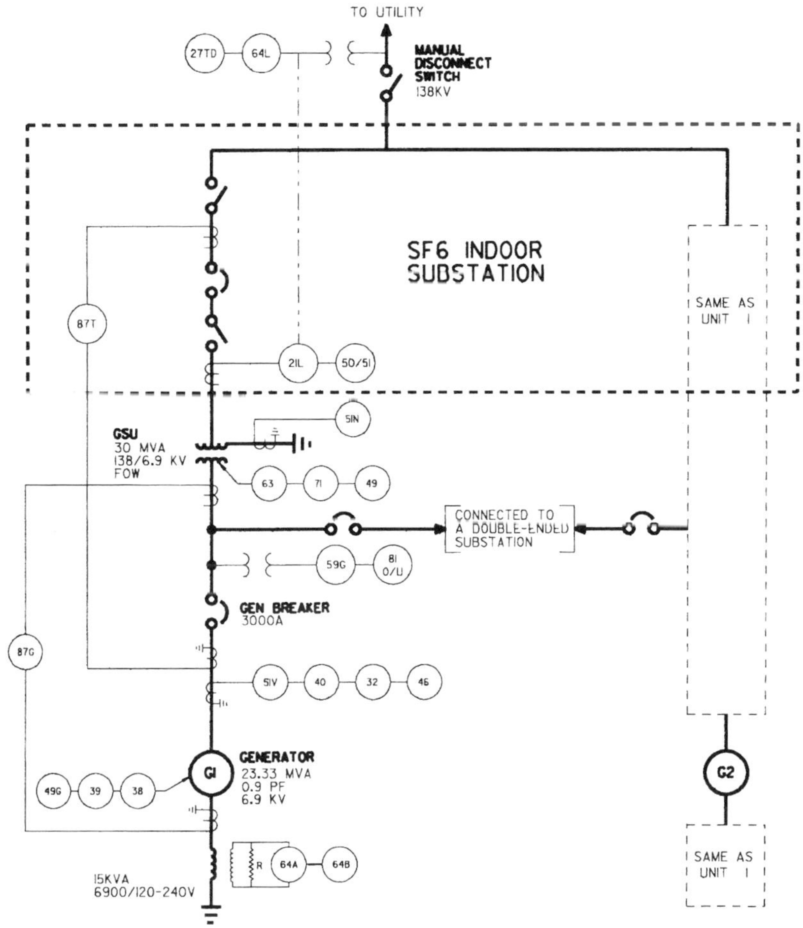

FIGURE 10-6: ***Protective Relay Scheme at the Belleville Hydroelectric Project.*** *These protective relays were installed in 1991 at an Ohio hydroelectric plant with two 23.33-MVA units.*

---

fully functional.

Individual solid-state and induction-disk type protective relay systems that use one relay for each protective function require more frequent testing and more extensive preventive maintenance. During testing, each protective relay must be physically removed from the switchboard on which it is located and placed on a bench or table next to the relay test equipment. Each relay is then tested to confirm the seal-in contacts, pick-up targets, time current characteristics, and restraint coils are operating properly for the individual protective settings of the relay. This is a time-consuming operation and requires test personnel knowledgeable in the specific types of protective relays installed at the plant.

REFERENCES

[1] "IEEE Guide for Control of Small Hydroelectric Plants," ANSI/IEEE Standard 1020, American National Standards Institute (ANSI)/Institute of Electrical and Electronics Engineers (IEEE), 1990.

[2] "IEEE Guide for Control of Hydroelectric Powerplants," ANSI/IEEE Standard 1010, ANSI/IEEE, 1991.

[3] "IEEE Recommended Practice for Protection and Coordination of Industrial and Commercial Power Systems," ANSI/IEEE Standard 242, ANSI/IEEE, 1986.

BIBLIOGRAPHY

Vadari, Subramian, Dennis Harding, and Larry Douglas, "Open Systems for the Electric Utility Industry: What, Why, and How?" IEEE *Transactions,* No. 0-7803-1301-1, 1993.

"Communication Protocols," IEEE Tutorial Course 95-TP-103, IEEE Press, New York, NY, February 1995.

Chapter 11
Testing Unit Control Switchboards In the Factory and Field

Chapter 11

At most of the smaller hydroelectric plants (up to 20 MW of capacity) built in the past 50 years, all the sensors and controls are connected to a unit control switchboard (UCS). In this way, the plant operator can control the entire station from one location. Since a malfunction in the UCS can hamper the entire hydro plant's operation, the switchboard needs to be highly reliable.

One way that owners of hydro plants can prevent UCS operating problems is to insist on a series of factory and field tests prior to startup and commissioning. These tests allow evaluation of the UCS design before it leaves the factory as well as a functional checkout when it first arrives in the field. Details regarding both factory and field tests should be outlined in the equipment specifications for the unit control switchboard. By performing tests in the factory and in the field, any problems with the switchboard will likely be identified and solved before unit start up and commissioning. By ensuring that the UCS is operating error-free during the testing phase, the startup and commissioning phase can proceed without costly delays for correcting a defect in the UCS design.

This chapter primarily addresses small to medium-sized hydropower plants, where a programmable logic controller (PLC) performs the control functions. Larger plants, where a distributed control system replaces the PLC, are addressed in a separate section at the end of the chapter.

Specifying the Tests

The unit control switchboard is the main operator control point from which unit start and stop operations are normally performed. The UCS usually contains a PLC and all associated plant recorders, meters, annunciators, and protective relays to give the operator complete control capability. Because of the variety of parts comprising the UCS, the risk of operating problems is substantial.

An owner or engineer responsible for purchase of a UCS should confirm that a detailed list of factory and field tests are included in the equipment specification. The factory tests will verify that the UCS was constructed properly at the factory. The field tests will confirm that the UCS was shipped and installed properly. Harza's experience has shown that specifying a detailed list of factory and field tests for major equipment at a small hydro station (including the UCS) can result in a startup and commissioning period of less than ten calendar days from the time of unit water-up.[1]

Identifying Necessary Tests

Identifying which tests to specify for a UCS is more difficult than for other items of electrical equipment such as generators and transformers. The Institute of Electrical and Electronics Engineers/American National Standards Institute (IEEE/ANSI) standards, which list these tests, are general in scope to accommodate a wide array of control switchboards.[2, 3, 4] The generality of the standards prohibits the detail necessary to define a "complete" switchboard testing procedure or sequence. As a result, confusion can arise concerning the factory and field tests for switchboards.

Failure of a hydro plant owner to recognize that a complete UCS checkout is not explicitly detailed by the IEEE/ANSI standards has led to many conflicts between clients and switchboard equipment manufacturers. These misunderstandings have also resulted in many lengthy delays in unit startup and commissioning.

The author's experience has led to a recommended set of test sequences and procedures for installing a UCS at a small hydro station. These sequences and procedures are also directly applicable to testing of unit control switchboards at large hydro stations. The basic recommended factory and field tests are outlined in Tables 11-1 and 11-2. These tables are not meant to be inclusive. Additional tests should be added as required, depending on the equipment supplied with the UCS for a specific project.

Because the UCS consists of many component and subcomponent systems, no single voltage test or wiring check will verify the integrity of the entire switchboard. Instead, a set of tests should be established by the owners to verify:

- The correctness of the internal wiring;
- The factory and field calibration and the correct operation of all sensing and recording devices;
- The proper operation of all control and instrumentation systems; and
- The proper operation of all protective relays.

For a UCS system, the functional test is the ultimate test; the proof is in performance.

TABLE 11-1: ***Factory Tests for Unit Control Switchboards***

1. Review front and rear elevations versus the final approved drawings. Check each item of equipment for proper location and verify the instrument/catalog number is correct per the specification.
2. Review the interior of the UCS in the same manner as the elevations. In addition, verify the lighting is adequate and grounding connections are provided.
3. Check anchor channels and cable entrances. Confirm they are in accordance with the drawings.
4. Review test certificate or witness the insulation resistance test of all wiring, current transformers, and potential transformers.
5. Check approximately 5 to 10 percent of the internal cabling. Verify that the following items conform to the drawings:
   - Cable numbers;
   - Terminal block designations;
   - Terminal designations on individual components such as control switches and lockout relays;
   - Raceway layouts; and
   - Equipment identification nameplates.
6. Activate all protective relays. Confirm that the appropriate lockout relay is energized and the correct annunciation and/or printout occurs.
7. Confirm that settings of all protective relays are in accordance with approved documents.
8. Check all annunciation points.
9. Check factory calibration of all devices possible, including electronic speed relays, current and potential transformers, and vibration monitors.
10. PLC checks:
    - Check the I/O racks for type and number of analog and digital I/O cards;
    - Check for future expansion capabilities on the I/O racks;
    - Check for surge protection provided on the I/O rack and I/O cards;
    - Identify grounding connections for the PLC and the I/O rack; determine whether chassis and logic grounds are the same or separate (this will affect the type and quantity of external surge protection required);
    - Review the PLC ladder diagram viewed on the video display terminal versus the final approved PLC software coding documentation; and
    - Verify that modem connections are provided and functional.
11. Perform the function checks listed below with the final approved schematics, PLC software coding, and control block logic diagrams in front of you. All permissives and interlocks should be provided by using the "dummy" toggle switchboard to provide these inputs.
    - Manual start/stop sequence (does not apply to redundant PLC control schemes);
    - Auto start/stop sequence;
    - Manual emergency stop sequence;
    - Automatic emergency stop sequence (usually performed by activating one of the lockout relays while in the "normal running" mode);
    - Change position of all control switches as follows (typically done while in the normal running mode):
      - local control to remote control
      - remote control to local control
      - manual control to automatic control
      - headwater level control "OFF" to "ON"
      - headwater level control "ON" to "OFF"
      - excitation manual control to excitation automatic control
      - excitation automatic control to excitation manual control; and
    - Verify the performance of the automatic synchronizing circuit and the manual sync-check relay (if provided).

Witnessing the Tests

It is imperative that the owner or engineer witness the factory testing of the UCS. The manufacturer should not be allowed to ship the unit control switchboard until factory tests have been completed with the owner or engineer as a witness. To ensure this policy, the "witness procedure" should be written into the equipment specification.

Factory testing, as demonstrated in Table 11-1, consists of a great number of drawing checks, calibrations, and functional checks. Complications during any of these sequences, although minor in the factory, could substantially affect the system in the field. Therefore, the owner or engineer must be present at the factory test to determine what is a minor problem that can be corrected in the field and what is a major problem that can't.

Moreover, most functional checks are not documented extensively—that is, a step-by-step account of the procedure is not detailed in the written test report. Therefore, it is mandatory that the owner and his personnel be present to note which systems failed the functional tests and exactly how those failures were remedied. This experience will prove valuable during the field functional tests because similar problems frequently reoccur in the field.

Importance of Factory Testing

Factory testing ensures that each individual system within the UCS has been tested and proved correct before the equipment is shipped to the site. Factory testing is beneficial to the owner for several reasons. The factory test occurs several months prior to field testing of the UCS and unit startup and commissioning. Therefore, if any major equipment errors or control system errors are detected during the factory test, the manufacturer has several months to solve the problem without affecting the unit commissioning date.

Correction of a problem in the field is more expensive than the same correction in the factory. If a problem develops or a component is found defective in the field, the minimum time required to receive a new component part is usually two days. In many cases, this component failure delays progress by the field test personnel until the new part is received and installed. Since the manufacturer's field labor is more expensive than factory labor, it is not only logical, but frugal, to test as many equipment components as possible in the factory before shipping the UCS to the site.

A number of the UCS components and systems lie directly in the critical path of unit startup and commissioning. If these components are not factory tested, their failure will delay the unit commissioning in direct relationship to the amount of time required to ship by air freight and install the new component part. A delay means the owner is paying for the UCS manufacturer's field labor and for the labor of associated startup and commissioning crews that are affected by this delay.

An example of this sequence of events involves the UCS synchronizing circuitry. If this circuitry is not factory tested, the owner's field personnel will not be able to functionally check the circuits in the field until the unit is watered up and the first synchronization to the utility system is attempted. Any failure of the synchronizing components at this time will be costly, especially if the project schedule is delayed.

Finally, during unit startup and commissioning, a significant number of unanticipated problems inevitably will occur. These problems will require the owner's undivided attention. Therefore, eliminating all "factory solvable" problems (problems internal to an individual item of equipment) will reduce potential delays in the project schedule. This is particularly applicable to the switchboard because of the quantity of sensors and control systems and subsystems involved.

Field Testing

The continuity of test experience from the factory to the field will expedite field testing of the UCS and, consequently, the startup and commissioning of the unit. The same test personnel utilized on the UCS manufacturer's test floor should be available at the site during checkout and field testing of the switchboard. This will provide continuity concerning problems or peculiarities associated with the various control systems, subsystems, and individual components. Such continuity is important because of the quantity and diversity of control and instrumentation systems connected to the UCS.

Recommended field tests for unit control switchboards are listed in Table 11-2.

Table 11-2: ***Field Tests for Unit Control Switchboards***

1. Verify tags on all factory-calibrated instrumentation devices.
2. Check all external interconnection wiring against the approved powerhouse/equipment drawings, verifying the following items:
   - Cable numbers and type;
   - Terminal block designations; and
   - Raceway layouts.
3. Perform point-to-point continuity and megger tests on all external cabling.
4. Calibrate all remaining instrumentation devices.
5. "Bench test" all protective relays to ensure proper settings.
6. Perform functional checks tests on all unit and station auxiliary equipment controlled from the UCS to verify proper operation.
7. Perform functional checks on unit start/stop sequences, duplicating the factory sequences listed in Table 11-1, step 11. These checks should be performed first with the associated power circuits de-energized, and then with both power and control circuits energized.
8. Methodically document steps 1 through 7 to ensure that no cables, instrumentation device, protective relays, or control systems have been overlooked.
9. Water-up the unit and perform all start/stop sequences listed in Table 11-1, step 11.

Documentation During Field Testing

Equipment specifications should identify the party responsible for the startup and commissioning of the unit (usually the turbine manufacturer in the case of a small hydro), and require that specific party to provide a coherent field checkout testing document. The document should include a standard checkout sheet for each control system. An example of a worksheet for one system is shown in Figure 11-1.

Moreover, the field checkout testing document should identify the sequence of testing for the various control systems related to the switchboard. This sequence of testing will allow the person in charge of startup and commissioning to coordinate the site visits of the manufacturers' representatives. For example, on most small hydro projects the UCS manufacturer usually provides one field representative to field test the UCS. This person can only check one system at a time. Therefore, it would not make sense to have the governor, generator, and intake gate field representatives expecting the UCS field representative to check out their interconnecting cabling on the same day.

During field checkout of the UCS, the following items should be documented:

- Identification of the control system or subsystem to be checked or tested;
- Verification that the associated interconnecting cables for this control system have been "point-by-point" continuity checked and megger-tested;
- Calibration of control system instruments;
- Functional check of control systems; and
- Verification (initials or signature) identifying the representatives who checked the system.

The field checkout documentation should be submitted to the owner for review and approval.

Field checkout documentation is essential for a methodical check of not only the UCS control systems, but also the various field component systems interconnected to the UCS, such as the governor oil pump system and its associated valves and solenoids. This checkout procedure will confirm that the UCS control systems have been field tested and are operating properly prior to the first synchronization of the unit. This should guarantee that the subsequent startup and commissioning of the unit proceeds smoothly.

Typical Problem Areas

Certain items of electric equipment have a tendency to create similar, if not identical, problems at each hydro site. Therefore, it is especially important to emphasize these items in the specification test section.

Automatic Synchronizing and Sync-Check Relay Circuitry

A factory test witnessed by the owner is usually the only time the automatic synchronizing and sync-check relay circuitry will be checked. This test requirement is

COMPLETED ✓

COMMISSIONING JOB WORKSHEET

HSPE PROJECT NO. C5082

SYSTEM NO./DESCRIPTION 7.0 - Generator & Plant Control System

TASK NO. 7.0.A/B TEAM Electrical SHT 3 OF 4

TASK STEP NO.	DESCRIPTION	INITIALS	DATE
11.	Verify all protective metering relays are installed in the correct location.	DGG	6/6/87
12.	Visually inspect relays for mechanical integrity.	DGG	6/6/87
13.	Verify all metering and protective relays have been field calibrated to HSPE specifications.	DGG	6/6/87
14.	Verify that the annunciator alarm and shutdown circuits have been functionally proved from the sensing device to the annunciator. The horn may be inhibited for these tests.		
15.	Verify gate position meter and lights indicate "CLOSED".	DGG	6/6/87
16.	Depress E-Stop push button to reset.	DGG	6/6/87
17.	Verify turbine case pressure positive indicator is "OFF" when case is empty. Lamp should light above 75 psig set point.	DGG	6/6/87
18.	Verify location and smooth operation of the following switches. Leave them in the "OFF" position.		
	o Speed limit	DGG	6/6/87
	o Gate limit	DGG	6/6/87
	o Generator brake	DGG	6/6/87
	o Voltage adjust	DGG	6/6/87
	o Bus volts		

FIGURE 11-1: ***Control System Commissioning Worksheet.*** *A commissioning job work sheet such as this should be used during field checkout to ensure that all electrical/mechanical systems have been tested prior to operation. Upon completion of the work sheet, it should be turned over to the owner or engineer for review and approval.*

often omitted from specifications, especially for small hydro stations. Plant owners need to be aware that they can write this test into their specifications.

To perform the test, the manufacturer must have access to an independent generation source. Many manufacturers will claim such a source is not available in their shops, and resist bringing in a separate motor-generator set. However, this test should be enforced if it has been written into the specification. It is actually not difficult to obtain a motor-generator set to perform the test. Checking circuitry in the factory will save the two days of field time required to ship by air freight and install a new sync-check relay if the relay malfunctions at the site.

Programmable Logic Controllers

Testing procedures for the PLC at both the factory and in the field should be clearly identified in the specifications.

Before the scheduled factory test, the owner or engineer should receive the final PLC software coding documentation for review. All review comments or changes to ladder logic should be implemented prior to the PLC's arrival on site. The owner should insist that at least one spare input/output (I/O) card of each type (analog, digital, RTD, etc.) is supplied with the switchboard when it is shipped to the site. Often at least one, if not more, of these I/O cards will fail during the UCS field checkout and subsequent commissioning phase. Having the spare card available will save unnecessary delays.

During the factory test, the owner should carefully scrutinize the skill of the person operating the CRT interface keyboard to ensure his or her capability. Insist on a replacement if the operator is inadequate. In addition, ensure that the same CRT operator working at the factory test will be at the site during field checkout of the PLC and unit startup and commissioning. The operator's ability to edit or modify the ladder logic and analyze the analog inputs and outputs for counts, alarm, and trip levels will save a number of days in the field during startup and commissioning.

Typically, numerous revisions are made to the PLC software coding document during the factory tests. The owner should ensure that he has received the final PLC coding documentation after the completion of the factory tests. The documentation should then be issued to the engineer and field personnel before the field checkout of the UCS. This document is essential for the functional operational tests of the control systems.

Once the PLC is at the site, the owner should pay particular attention to its setup. It is important to review the surge protection provided for the PLC and associated hardware. Transient voltage surges are a major problem with PLCs, especially during the summer thunderstorm season. Grounding connections also require investigation. The amount of surge protection required for the PLC hardware components depends upon the grounding connections recommended by the PLC manufacturer. Many PLC manufacturers recommend connecting their chassis (frame) ground to their logic ground. Some manufacturers let their logic ground float. For an electrically noisy environment like a powerhouse, allowing the logic ground to float is preferable. The owner should be aware that the amount of surge protection required is related to the grounding method adopted.

At the end of each day during startup and commissioning, the engineer

FIGURE 11-2: *Dillon Dam (Colorado) Control System.* *The 2-MVA Dillon project is used for flow control and power generation. This photograph of the unit control switchboard shows two panels on the far left that house the unit controls. The protective relays are mounted on the middle panels. At the owner's request, the control scheme utilizes a complete hardwired relay system. A UCS associated with a hardwired relay system still requires testing.*

should make a cassette tape or floppy disk hard copy of all software changes of the PLC coding during the day. If a PLC battery malfunctions or another type of PLC outage occurs, this backup hard copy can be used, and thus startup delays can be avoided.

Considerations at Large Hydro Plants

The preceding section was written to address small and medium sized hydro plants that use PLCs for automatic operation of the hydro plant. It does not address computer-based SCADA systems (for example, distributed control systems or DCSs), which are installed at large, multiple-unit hydro plants. However, the testing sequence (factory tests prior to shipment and field tests after installation) is still valid. Both PLC and computer-based SCADA systems must have a well-defined equipment specification that details the factory and field tests to be performed.

In general, due to the more complex operating system with a computer-based SCADA system, a number of additional system functions should be tested. These tests are listed in Table 11-3.

TABLE 11-3: ***Additional Factory and Field Tests for Distributed Control Systems***

1. Point-by-point database check.
2. Database linkage to graphical displays.
3. Response times during normal loading and high activity loading scenarios for:
   - Graphical display updates;
   - Control sequence implementation;
   - Alarm processing and logging; and
   - Sequence of events recording.
4. Communications connectivity/protocols.
5. Man-machine interface (MMI) user capabilities.
6. Application software functionality.

The factory tests of the control system should verify that the equipment was constructed properly at the factory. The factory tests should verify that the control/computer system meets all the specification requirements concerning response time, communication protocols, software configuration, and hardware components. The factory tests provide an important baseline for the field performance tests of the system, which are performed at the site during start-up and commissioning of the plant.

The field tests will confirm that the control/computer system was shipped without damage, was installed properly at the plant, and can duplicate the response times and other performance measures achieved during the factory tests. Harza's experience has shown that specifying a detailed list of factory and field tests for all plant control/computer systems, including the small hydro station PLC control systems, results in early discovery of any misunderstandings concerning the specification requirements. This early discovery allows the appropriate corrections to the hardware and software to be implemented at the factory. This results in a reduced start-up and commissioning time period for the equipment in the field.

REFERENCES

[1] "Start-Up and Commissioning Procedures for Hydroelectric Units," *Hydro Review*, Vol. 8, No. 3, June 1989, p. 64-70.

[2] "IEEE Guide for Control of Small Hydroelectric Plants," ANSI/IEEE Standard 1020, American National Standards Institute (ANSI)/Institute of Electrical and Electronics Engineers (IEEE), 1990.

[3] "IEEE Guide for Control of Hydroelectric Powerplants," ANSI/IEEE Standard 1010, ANSI/IEEE, 1991.

[4] "IEEE Standard for Control Switchboards," ANSI/IEEE Standard C37.21, ANSI/IEEE, 1985.

Defining Testing Terms

A variety of terms will arise during factory and field testing of the unit control switchboard. Varying interpretations of these terms create a serious potential for miscommunication between owner and manufacturer. The following is a description of some of the most frequently used testing terms.

Functional Test: Test in which the various control switch, permissive, and interlock contacts illustrated on the unit control schematic diagrams and the PLC ladder are sequentially "functioned" to prove that the logic is correct.

Functional tests are usually proved on a system-by-system basis. For example, the functional test for the governor oil pumps would first prove manual operation of each pump is possible, followed by automatic operation (including the lead/lag function). Then the test would verify that all pump control indications and interlocks are functioning as shown on the ladder logic.

Field Checkout Documentation: Documentation that certifies the various control systems were tested and functionally checked to verify their operation. Figure 11-1 shows an example of this documentation.

Several contractors are involved in providing the field cabling and connections to the unit control switchboard. Therefore, the documentation should identify which contractor is responsible for assisting the UCS field representative in performing the continuity, megger, and functional tests on the various systems.

The field checkout testing of the switchboard precedes the actual synchronizing of the unit and should ensure that the commissioning of the unit proceeds smoothly.

Bench Tests of Protective Relays: Field tests of the protective relays. The relays are physically removed from the switchboard and placed on a bench or table. The relays are then tested with the appropriate test equipment to verify that they function properly at the approved settings. Seal-in contacts, pick-up targets, time current characteristics, and restraint coils are also verified during this test procedure. Bench tests of microprocessor-based multifunction relays are performed with the relay remaining in place on the switchboard. Communication connections are made between the relay and a personal computer (PC). Software installed on the PC allows the PC to interrogate and calibrate the relay.

"Dummy" Toggle Switch Board: Board, usually made of plywood, on which various toggle switches are mounted and labeled according to the permissive or interlock they represent. The board is utilized during the factory test of the switchboard to send fake or "dummy" inputs to the PLC for devices not physically present in the factory. The dummy inputs could represent a wicket gate speed-no-load position switch, a cooling water flow switch, or a penstock inlet valve open switch. All these switch inputs would be required as permissives to complete the unit automatic start sequence.

Chapter 12
Surge Protection

Protection from lightning and other transient electrical surges is a critical part of design of electrical and control facilities for a hydropower plant. This is increasingly true as hard-wired control systems are replaced by solid-state digital components, which are more vulnerable to damage from electrical surges.

Direct Lightning Protection

Design of lightning protection systems requires a basic understanding of the physics of lightning and the means available to both divert and conduct lightning stroke energy around structures. Once the basics are understood, an engineer can use existing codes and standards to design an effective protection scheme for a given structure. A well-implemented scheme ensures that direct or close proximity lightning strokes will not cause forced outages of hydroelectric generator units. Lightning protection should be a part of the design of any new hydroelectric plant. If protection was not provided when the project was constructed, rehabilitation projects and activities offer opportunities to make corrections.

Reviewing Lightning Basics

Originally thought to be a direct current (DC), lightning is a transient high current electric discharge consisting of an initial discharge with multiple return strokes. Most frequently the initial discharge occurs because of the separation of charged particles within a cumulonimbus cloud. (A cumulonimbus cloud is a dense cloud developing vertically through all cloud levels, consisting of water droplets, ice crystals, and sometimes hail, and associated with thunder, lightning, and heavy showers.) Two major theories exist about how the positive and negative ions form and accumulate. The most widely accepted theory contends that precipitation formation solely within the cloud causes the ionic separation, with ice particles (positive charges) moving to the top and water particles (negative charges) moving to the bottom. The other theory states that the charges are caused by convected updrafts and downdrafts. The warm updrafts carry positive ions generated from sharp points on the ground to the top of the cloud. Cold downdrafts carry negative ions (as water particles) to the bottom of the cloud.

The initial discharge for most lightning strokes (90 percent) occurs when the accumulated negative charge in a cloud discharges to the ground. However, a certain percentage of lightning is from cloud positive to earth. The type of discharge depends on the cloud structure and how the positive and negative charges within the cloud arrange themselves. Upward (or ground-initiated) strokes rarely occur. However, this type of stroke can occur from mountaintops or other tall manmade structures (for example, transmission line towers). A good explanatory reference concerning types of lightning can be found in the *International Geophysics Series*.[1] An event catalog titled *Lightning, Auroras, Nocturnal Lights, and Related Luminous Phenomena* details specific historical occurrences.[2]

Because of the number of return strokes between the cloud and ground after the initial discharge, lightning is, in effect, a high frequency alternating current (AC). The number of return strokes varies depending on the cloud's formation and altitude, the air temperature, and subsequent leader channel resistance. An initial stroke usually reaches its peak in microseconds. Return strokes occur 10 to 100 milliseconds after the preceding stroke. The first stroke usually has the largest current and can range from 10,000 to 100,000 amperes.[3] Lightning discharges generate a broad range of frequencies from 10 kHz to above 10 MHz. The rise times of the initial stroke and subsequent return strokes determine the spectrum of the energy discharge.

Lightning can be classified as either cold or hot. Cold lightning has a short time duration (approximately 1,000 microseconds from the beginning to the end of the initial stroke). When cold lightning strikes objects, a physical reaction occurs — usually an explosive one. For example, when cold lightning strikes certain trees, it causes a rapid vaporization of the water content immediately inside of the bark. This results in a piece of bark "blowing off" as the stroke progresses toward the ground. Hot lightning has high currents, a much longer time duration (approximately 2,000 to 200,000 microseconds), and a tendency to start fires.

The peak atmospheric channel temperature for both hot and cold lightning is approximately 50,000 degrees Fahrenheit (28,000 degrees Celsius).[3] (This rapid heating of atmospheric gases causes thunder.)

Predicting Lightning Activity

The amount of lightning that strikes a geographical region is a function of the number of thunderstorms that occur. Until the mid-1990s, utilities typically estimated lightning frequency from the U.S. Weather Bureau's "isokeraunic" maps of the U. S. On these maps, isokeraunic lines (representing the number of days on which thunder could be heard) were established from U.S. Weather Bureau records of 266 weather stations. For instance, the isokeraunic lines show that a location in Georgia will normally have between 60 to 70 thunderstorms per year, whereas a site in Iowa will have between 40 to 50 thunderstorms per year.

Some utilities still use the isokeraunic map to identify the lightning activity in their area, and hence the level of surge protection needed for their transmission and distribution lines. However, the isokeraunic map gives only a rough estimate of thunderstorm activity, primarily due to the low density of weather stations over the large area of the contiguous U. S. (approximately 3,600,000 square miles) and to the fact that thunder cannot usually be heard more than 15 miles from the originating lightning stroke.

In recent years, the National Lightning Detection Network (NLDN) has

Glossary of Lightning Protection Terms

Definitions are extracted in part or in their entirety from American National Standards Institute/National Fire Protection Association Standard 78 and modified, where necessary, for additional clarity.

Air Terminal: Pointed solid or tubular rods of specified size and material provided with a mounting base and having a proper conductor connection. The terminal is commonly called the "Franklin air terminal" after Benjamin Franklin. This nomenclature distinguishes the Franklin air terminal from specially designed air terminals used in "lightning attraction" or "dissipation" systems.

Bonding: A connection between a metal object and an element of a lightning protection system to accomplish electrical continuity between the two.

Down Conductor: The portion of a lightning protection system designed to carry the lightning discharge between the air terminals and ground.

Ground Mat: An interconnected system of ground rods and ground conductors located within or close to a structure of building. The ground mat connects equipment to earth potential.

Leader Channel Resistance: The atmospheric resistance that the initial positive or negative ionized streamer encounters. This streamer establishes the channel, or atmospheric path, in which the lightning current flows.

Primary Metal Body: A metal object on or above roof level that is subject to a direct lightning stroke.

Secondary Metal Body: A metal object at or below roof level that may build up a charge that is opposite to the ground or to the lightning protection system.

Side Flash: A spark occurring between a metallic object and the lightning protection system, the ground, or another nearby metallic object.

Zone of Protection: The adjacent space to a grounded air terminal, mast, or overhead ground wire that is substantially immune to direct strokes of lightning.

FIGURE 12-1: ***Franklin Air Terminals.*** *Lightning rods, as seen in this photograph atop the two transmission towers at the 115-kV Moose River Substation, are commonly known as Franklin air terminals. An area around each air terminal, called the zone of protection, is substantially immune from direct strokes because the air terminal attracts and receives the lightning stroke instead of lower objects in the immediate vicinity. Various heights of air terminals may be selected, depending on the zone of protection required.*

become accepted by utilities as a lightning resource facility. The NLDN was originally developed by the State University of New York under Electric Power Research Institute (EPRI) sponsorship, and is presently operated by Global Atmospherics, Inc., as a private enterprise.[4] NLDN has over 106 strategically located sensors throughout the continental U. S. which allow its operators to physically locate 90% of the cloud to-ground lightning strokes that occur in the country. The NLDN can also identify the peak currents associated with each lightning stroke, the polarity of the stroke, and the number of strokes. The NLDN compiles historical flash density maps for different utility areas. These maps allow utilities to identify the level of surge protection that is required for a given transmission line. (The level of surge protection required for a specific transmission line is determined by a PC software program designed by the Electric Power Research Institute called "Lightning Protection Design Workstation."[4])

In addition, many utilities use the NLDN data to alert repair crews and prepare for high intensity lightning storms in their area. NLDN is able to furnish real-time lightning occurrences to subscribers by a satellite link. Consequently, the utility subscribers are able to identify whether an incoming thunderstorm is a high intensity lightning storm, and dispatch personnel and equipment appropriately. Documentation from the NLDN system also assists in resolving insurance claims by proving the time and location of specific lightning strikes.

Schemes for Lightning Protection

For protection from direct lightning strikes, a structure needs a mechanism to prevent the lightning from striking the structure itself and a way to channel and dissipate the lightning's energy. The focus of a direct lightning strike protection scheme is to divert the discharge around the power structures to dissipate in the ground mat.

Protective Devices

Lightning rods, commonly known as Franklin air terminals (or simply air terminals), protect structures from direct lightning strokes. An area around each air terminal, called the zone of protection, is substantially immune from direct strokes because the air terminal attracts and receives the lightning stroke instead of lower objects in the immediate vicinity. When lightning hits, the air terminal and its associated connections convey the lightning current safely to the ground grid of the structure. The American National Standards Institute/National Fire Protection Association (ANSI/NFPA) Standard 78 defines air terminal dimensions, air terminal quantities, size and quantities of down conductor cables, fastening methods and other specifications for air terminals.[5]

To avoid induced voltage side flashes, ANSI/NFPA 78 suggests bonding any other steel structures on the roof to the lightning protection ground grid. Figure

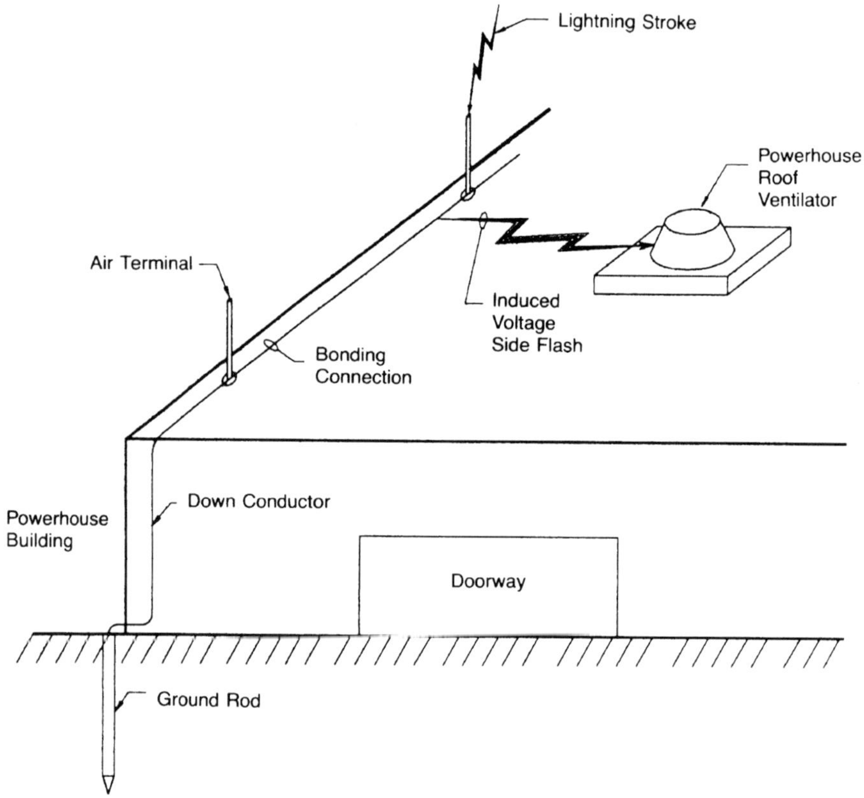

FIGURE 12-2: ***Lightning Protection System.*** *A typical lightning stroke attempts to push 25,000 amperes of current to ground within a few microseconds. This produces a large induced voltage side flash in any metal object-specifically powerhouse roof ventilators or other ungrounded metal structures located on roofs with the possibility of side flash, as illustrated in this figure. Therefore, all metal objects should be bonded into a single grid type system that is routed to ground.*

12-2 shows a single grid type system, in which all metal objects (except the powerhouse roof ventilator) are bonded into one system, which is routed to the ground.

In general, the radius of the zone of protection of air terminals is determined based on experience; however, codes in various countries differ as to the exact radius that should be protected.[6] Before 1989, ANSI/NFPA 78 stated that an air terminal will protect a conical volume with a base dimension of 2H for a vertical air terminal height of H (ratio of 2:1). Other countries, such as Canada and England, used a ratio of 1:1. These ratios are for simple structures. For more complicated structures, such as an intake gate or spillway gate at a hydroelectric

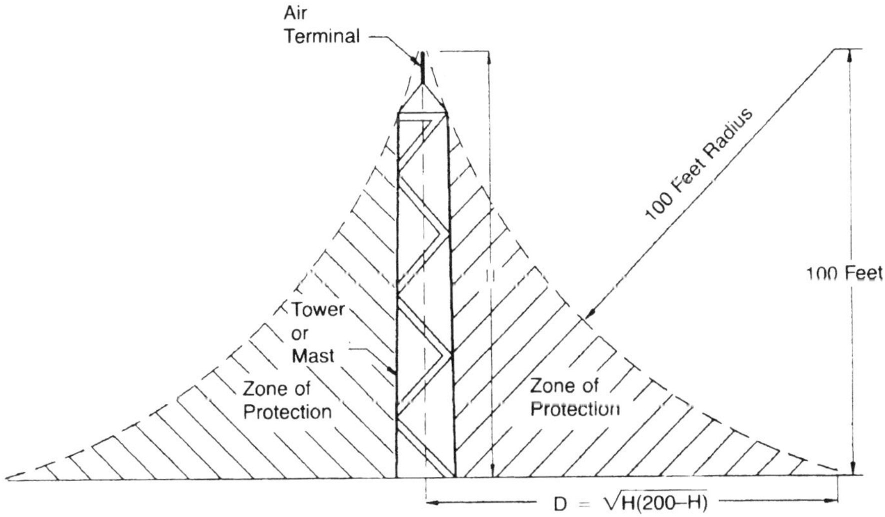

FIGURE 12-3: ***Zone of Protection.*** *This figure illustrates the zone of protection for an air terminal, according to paragraph 6.3.3 of the American National Standards Institute/National Fire Protection Association Standard 78 (1989 revision). The arc, inscribed from the tip of the air terminal to the ground plane, has a maximum radius of 100 feet. The protective zone radius is based upon experience and observation. Different countries specify different protective zone radii for acceptable protection. In the equation given, D is the distance from the air terminal to the edge of the inscribed arc, and H is the height of the air terminal (less than or equal to 100 feet).*

---

plant, ANSI/NFPA 78 defines the exact quantity and locations of air terminals with detailed figures and descriptions. In 1989, owing to better research and data, ANSI/NFPA 78 was updated. As part of the update, the zone of protection was redefined. In Figure 12-3, the zone of protection is the volume under an arc inscribed from the tip of the air terminal to the ground with a maximum radius of 100 feet. Lightning protection experts encourage reference of the ANSI/NFPA 78 specifications to ensure good engineering practices for protecting hydroelectric power plants, associated intake gatehouses, and any endangered outlying buildings or structures.[7]

Besides passive Franklin air terminal systems, two other lightning protection schemes for buildings exist: lightning attraction systems and dissipation systems. Lightning attraction systems consist of spherical or saucer-shaped air terminals in place of the standard Franklin air terminal. The shape of the air terminal generates a stream of ions that attracts the lightning stroke in the same way as a naturally occurring upward or ground-initiated stroke.[8] Lightning attraction systems are used if a building or other location has a significant amount of very expensive electronic equipment. These types of systems are often used by the U.S. Federal

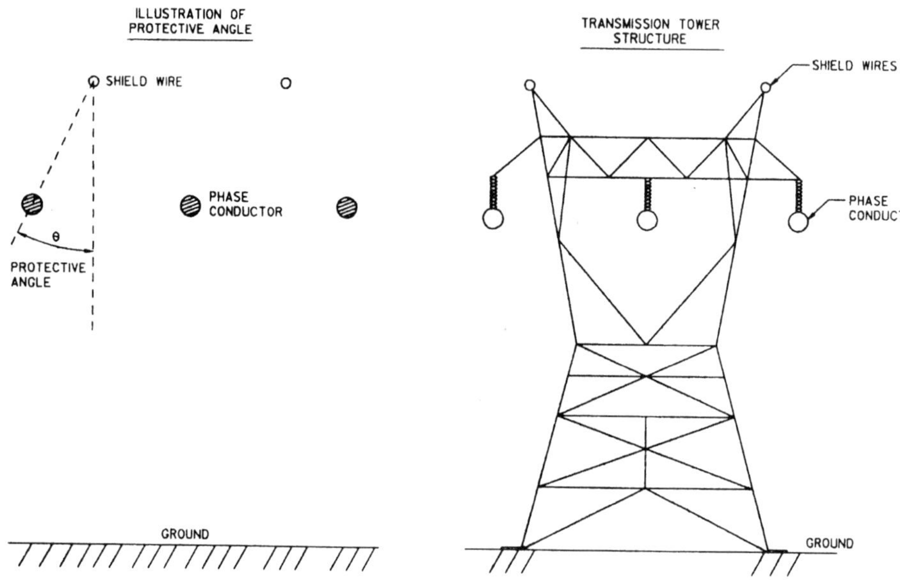

FIGURE 12-4: *Shield Wires.* *Transmission shield wires protect power lines by intercepting direct lightning strikes before they reach the phase conductors on transmission towers. The location of the phase conductor with respect to the shield wire forms the protective angle. The taller the tower, the smaller the protective angle needed for providing a minimum standard of protection.*

Aviation Agency and the Department of Defense. Dissipation systems consist of a multitude of sharp points rather than individual Franklin air terminals. The sharp points continually discharge the electrical field created in the earth during a thunderstorm.

Shield Wires

A shield wire (ground wire, static shield wire, or earth-wire shielding) performs the same function for a transmission line or a substation as an air terminal arrangement does for a powerhouse. Shield wires are above the 60-Hz phase conductors in a transmission line and, therefore, are physically closer to the cloud from which the lightning stroke originates. Shield wires protect transmission lines and substations by intercepting direct lightning strokes before they reach the 60-Hz power conductors or equipment.

Since transmission voltages have increased significantly over the past 50 years, modern power transmission lines require taller tower structures with wider crossarms. These taller transmission structures present a larger target for the lightning stroke and, consequently, are susceptible to an increased number of lightning strikes. Because of this, the location of the shield wire is a subject of extensive discussion. The discussion focuses upon the protective angle (shown in Figure 12-4), and thus the location of the phase conductor with respect to the shield wire. The taller the tower, the smaller the protective angle needed to provide a minimum standard of protection. Transmission tower heights of 50, 100, and 150 feet use protective angles of 45, 30, and 12 degrees respectively.[5]

Substations also can use shield wire arrangements for protection. In general, substation shield wires are connected directly to the substation ground mat. Three methods for direct lightning protection using shield wires exist from the 1940s and 1950s. Wagner's 1942 method, the fixed shielding angle method, and the rolling sphere method. A fourth method, Mousa's software subshield method, is more recent and is becoming more common.

To focus on one method, the Institute of Electrical and Electronics Engineers (IEEE) published Wagner's 1942 Method. Engineers use this method today with minor modifications incorporated in 1957.[9,10] It specifies the quantity, location, and height of substation shield wires required to establish an exposure probability of 0.1 percent. A 0.1 percent exposure probability for one substation allows only one direct stroke through the shield wires for every 2,000 to 4,000 years of operation.[9] As the number of substations increases, the overall exposure probability of a lightning strike on an individual substation increases in direct proportion.

The IEEE published a guide for direct lightning protection of substations in 1996, entitled "IEEE Guide for Direct Lightning Stroke Shielding of Substations." [11] This guide (IEEE 998) was developed to prevent confusion over the different methods utilized to protect a substation. IEEE 998 is a consolidation of the previously existing methods of direct lightning protection of substations and is now the definitive standard on substation shielding.

Debating Alternative Air Terminal Systems

Although passive Franklin air terminal systems are explicitly detailed in ANSI/NFPA 78, the effectiveness of lightning attraction and dissipation systems continues to be debated. In an NFPA standards meeting in July, 1995, the NFPA decided not to issue the pending Standard NFPA 781, "Standard for Lightning Protection Using Early Streamer Emission Terminals."[12] The unreleased standard defined a lightning attraction system that utilized special lightning rods that would emit a stream of ions that would trigger a lightning stroke. This would allow a structure to be protected by a reduced amount of air terminals.

After external review of the standard, the NFPA decided that it could not quantify the effectiveness of the lightning attraction system in comparison to a standard passive system and could not determine if the greatly reduced quantity of air terminals used to protect a structure in the attraction system was justified. This uncertainty resulted in a rejection of the proposed standard.

DETERMINING GROUND RESISTANCE

Standards for Ground Resistance

The Underwriters Laboratories' document UL 96A, "Master Labelled Lightning Protection Systems," specifies a "ball park" ground resistance value of up to 50 ohms for the ground rods connected to a specific down conductor.[13] On the other hand, ANSI/NFPA 78 does not specify a minimum ground resistance value. Instead, Chapter 3 of the standard, "Protection for Ordinary Structures," covers the following issues:

- Quantity and location of ground rods for different types of soil conditions observed;
- Installation connection methods to fasten the down conductors to the ground rods;
- Bonding together all of the ground systems of all primary and secondary metal bodies;
- Determination of effective bonding distances; and
- Determination of resistive and inductive components for grounded metallic bodies.

Following the installation techniques given obtains the "best achievable" ground connection, and, therefore, the minimum ground resistance possible for a specific soil type-regardless of the actual soil ohmic resistance. This can be confusing for power engineers who are taught that the soil ohmic resistance is a critical value in any powerhouse ground system design. For instance, ANSI C114.1/ IEEE 142 (the "Green Book"), which is specifically applicable to 60-Hz power systems, identifies an acceptable ground resistance of 1 ohm or less for large substations and generating stations, and less than 5 ohms for smaller substations.[14]

Test Procedures

The normal tests to determine an exact ohmic value for a powerhouse ground mat use low frequency (approximately 60-Hz) equipment. This test consists of either a three- or four-rod "fall-of-potential" test procedure. A test device generates a 60-Hz current (or slightly above) to measure the potential drop between the rods. Precise measurements of ground resistance are difficult to obtain even with present-day equipment. These measurements involve many indeterminate factors including various soil layers, last precipitation and amount of moisture retained in the various soil layers, and the temperature of the soil. The 60-Hz ohmic value determined takes into account ground electrodes and ground system connections (for example, powerhouse to switchyard) at large distances from a specific air terminal's down conductor connection point to the ground mat.

Although a low total ground resistance is desirable for the powerhouse ground mat (and essential for 60-Hz faults) this is not the same resistance (mainly impedance) that a lightning discharge sees. The high frequencies involved with most lightning discharges tend to move the discharge current to ground in the immediate area where the stroke occurs (the nearest down conductor and ground rods) because of the inductive reactances involved with any grounding cables more than six meters away. Therefore, the 60-Hz ohmic value determined by the normal ground system tests is only relatively indicative of the ground value a lightning discharge sees in the immediate strike area.

To test for lightning protection ground values, the engineer should isolate the ground rods connected to air terminal down conductors, and then test these rods for their 60-Hz ohmic values to obtain a ground resistance value. This value is more representative of the ground resistance value that a lightning discharge sees. However, since most powerhouse ground grids consist of an interconnected (cadwelded) ground system built during installation, this type of isolated testing usually is impossible, and in most cases is not required. In general, the powerhouse ground mat works for lightning discharges because a low 60-Hz ground resistance value usually indicates a corresponding low ohmic value (relatively speaking) for the lightning discharge. However, it should be emphasized that ANSI/NFPA 78 considers installation methods to be of greater importance than a 60-Hz ground resistance value for lightning protection systems.

Protecting Hydro Plant Control Equipment From Surges

Control systems in hydro plants have changed dramatically over the past twenty years. Mechanical-hydraulic or electro-hydraulic governors, hard-wired switches, and iron-vane meters have given way to sophisticated digital devices with solid-state components. Instead of the hard-wired relays of the past, today's control room uses programmable logic controllers (PLCs). Other solid-state devices provide inputs to or receive outputs from the PLCs (Figure 12-5). While this technology brings significant benefits to operational control, it is also more vulnerable to damage from transient electrical surges, such as those caused by lightning.

Modern solid-state components require that the plant owner and engineer investigate the need for expanded protection from transient surges. To design appropriate protection, the engineer must fully understand two concepts: the close interrelationship between the surge protection and the grounding system, and the effects of the higher frequencies that accompany transient surge voltages. A design that accommodates these concepts is needed to maximize the plant's on-line availability.

Standards for Surge Protection

There are numerous standards for surge protection, grounding, and acceptable installation procedures. A listing of these standards appears in this chapter's bibliography.

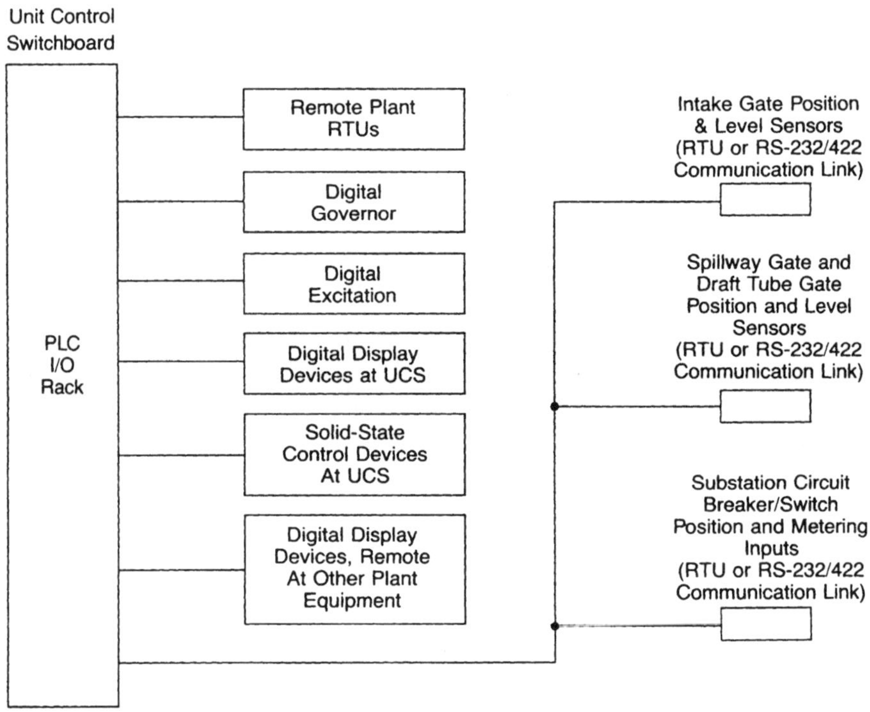

FIGURE 12-5: ***Hydro Plant Control System.*** *This schematic diagram highlights the types of control devices in most present-day hydropower plants. These control components are sensitive to transient surges.*

The most important recent document addressing these subjects is IEEE Standard 1100, (Emerald Book), "Recommended Practice For Powering And Grounding Sensitive Electronic Equipment."[15] IEEE 1100 provides a definition of terms inconsistently used throughout the industry, defines grounding criteria and power quality, provides case histories, and identifies recommended design/installation practices. This standard is a very important addition to the available literature.

Looking Beyond the Standards

Although the list of applicable standards is extensive, engineers should be aware of the subjects that some of these standards do not emphasize. For example, the emphasis in the *National Electrical Code* (NEC) is personnel safety. The NEC does not address the grounding requirements for proper operation and protection of control systems, especially the recently introduced solid-state component systems. A plant can be wired entirely in compliance with the NEC, and still experience control system failures caused by transient voltage surges.

Fortunately, the PLCs, input/output (I/O) racks, and solid-state components normally used in hydro plants have been industrially "hardened" to tolerate some amount of transient over-voltage. In an electrically "noisy" environment such as a powerhouse, these control components will continue operating after sustaining a surge that is within design limits. For example, most PLC central processing units are optically coupled to the I/O racks, and the I/O cards themselves have a specific over-voltage range that they can tolerate. Still, to ensure that lightning surges cause no damage, additional front end protection is required for certain I/O cards and for the PLC power supply. Although this additional surge protection can be retrofitted, providing it during initial installation of the control system may avoid unnecessary equipment damage.

Chapter 12

Methods of Surge Protection

The average lightning stroke consists of a 35,000-A current, with three to four strokes in a 20 to 100 millisecond time period.[3] The stroke is classified as high-frequency alternating current because the initial downward (or upward) stroke can generate as many as 20 responding stroke sequences.

The microsecond rise times of a lightning stroke produce frequencies in the MHz range. The hydroelectric plant owner and engineer must protect control devices from the MHz frequencies and their associated potentials, not the 60-Hz frequencies that are so familiar to professionals in the field.

The traditional protection from lightning is the plant's surge arrestor and its associated ground system, which provides the path to earth for the lightning current. IEEE 142 recommends that an acceptable value for ground resistance for generating stations and for large substations is less than 1 ohm; and for smaller substations and industrial plants, less than 5 ohms.[14] Generally, design engineers use a value of less than 5 ohms for small hydroelectric power stations (up to 20 MW) and less than 1 ohm for large stations. While this satisfies IEEE 142 and the NEC safety requirements, it is not low enough to eliminate electrical "noise" that can affect solid-state components. The RS-232 ASCII serial-communication circuits that are connected to the PLC operate at signal voltages ranging from +15 V for logic "zeroes" to -15 V for logic "ones." Any inductive voltages greater than +15 V will mask the RS-232 signal voltages and garble the signal. Such inductive voltages may be created by ground-loop currents from chassis to chassis, or from a remote transmitter to a powerhouse PLC.

Ground Systems — Historical and Current

In the early twentieth century, low-voltage AC power systems were ungrounded. This resulted in a high availability system at the expense of numerous personnel hazards. When a ground fault did develop on the system (for example at an equipment enclosure) the system would continue to operate; however, the person or persons who accidentally touched this enclosure while the ground fault existed risked severe harm or death. This system also resulted in numerous malfunctions of circuit breakers due to the floating system, in which breaker operation was dependent on the quality of the second ground developed. Consequently, the United States moved to a low voltage AC system with one live wire and one neutral wire, with the neutral wire connected to ground at the source. An important change to the system was to ground all equipment enclosures, handrails, and other structures which could come in direct contact with operating personnel. This eliminated the personnel hazards and gave the circuit breakers a high operational reliability at a slight expense to the system availability. This is the same AC system we have today.

The ground system is intimately related to the lightning protection system because the ground system provides the path to earth for the lightning current. IEEE 142 lists recommended acceptable values for ground resistance for large substations and generating stations as less than one ohm; and for smaller substations and industrial plants as less than five ohms.[14] The ground resistance values are applicable to 60-Hz systems, and are field tested with a low frequency (approximately 60-Hz) test set. These resistance values increase due to the high frequencies involved with lightning. The high frequencies create large voltage differential between equipment components.

IEEE 142 does not expressly discuss surge protection of control systems but was updated in 1991 to include a chapter on "Sensitive Electronic Equipment Grounding." The new chapter addresses methods of grounding to avoid voltage differentials between solid state/computer equipment components. The chapter emphasizes the use of a "single point method of grounding" for computers/controls versus a separate or isolated grounding method.[14] IEEE 1100 (the "Emerald Book") does address voltage differentials between equipment components and surge protection methods.[15] Consequently, the two IEEE standards should be used together to implement surge protection for solid state control systems.

Grounding the hydroelectric plant in accordance with both the NEC for personnel safety and IEEE 142 to obtain a one-ohm ground resistance does not eliminate transient surges that can penetrate the control system. Although many standards define transient-surge wave forms for testing the various surge protection devices, they do not specify minimum ground resistance or the impedance values associated with these wave forms.

Electrical and electronics engineers use three techniques to protect solid-state control equipment:

- Provide a low-impedance, single-point grounding system with the shortest possible connections to the equipment;
- Provide surge protection to eliminate or minimize all transient surge voltages to the equipment; and

- Interconnect the grounds for the powerhouse and switchyard mats. This interconnection is necessary because the two mats are already connected by various electrical pathways, such as the current transformer/potential transformer (CT/PT) circuits and instrumentation cable shield connections. Solid bonding between the two ground mats minimizes the impedance between these two systems, thereby minimizing the voltage difference that will occur during a lightning stroke.[16] If the individual ground mats are not solidly bonded, any surge voltage that occurs in one ground mat will find its way to the other via a higher-impedance path, creating a high transient surge voltage at the second mat.

Single-Point, Low-Impedance Grounding Systems

Single-point grounding systems connect all solid-state equipment grounds to each other and to a single-point, low-impedance earth ground. If a transient surge voltage occurs, this method of grounding raises all the connected solid-state equipment to the same potential level, preventing circulating loop currents between individual items.

Grounding systems have frequency-related problems. A grounding system may be a single-point system at 60 Hz, but at higher frequencies, stray inductive and capacitive couplings cause it to become a multiple-point grounding system (see box, "Relationship Between Frequency and the Impedance of a Ground Conductor.")

The higher the frequency, the more the stray couplings cause the 60-Hz, single-point ground system to react like a multiple-point ground system. This creates higher impedances, which result in differential voltages between various solid-state components. Once these differential ground-loop voltages exceed the component's signal voltage, errors in data transmission and even (depending on the magnitude of the voltage) equipment failure can occur.

To obtain a low-impedance ground at any frequency, the length of the ground conductor must be limited to one-half the wave length of the highest frequency for which the system is protected.[16] For example, to calculate protection against 30-MHz frequencies, the formula

$$(f)(\text{wave length}) = 3.0 \times 10^8 \text{ meters}$$

where 3.0×10^8 meters per second is the speed of light, gives a wave length of 10 meters, and consequently the ground conductor must be limited to one-half this distance, or 5 meters. This equation is similar to equations used for mechanical systems in which the speed of a wave travelling through a medium is determined by the properties of the medium. Consequently, once the speed of the wave and the frequency are determined, the wavelength can be derived.

The impedance of a ground conductor can be decreased by fabricating it out of large, braided-copper straps. The multiple pathways available to current flow will diminish the consequences of the skin effect at higher frequencies. Skin effect causes high frequency alternating current to flow on the surface of the conductor. The higher the frequency, the less the amount of conductor area used. Effective AC resistance related to frequency can be expressed as

$$R_{ac} = (k)(r_{dc})$$

where R_{ac} and r_{dc} are the resistance of the conductor to alternating currents and direct currents respectivley, and k has a specific value depending on the conductor size at a frequency of 1 MHz. Consequently, a 4/0 AWG cable at 1 MHz has a resistance approximately equal to an 18 AWG cable at direct current.

Minimizing Transient Surge Voltages

Although chassis grounds and logic grounds (electronic component grounds) are connected in hydroelectric plants, the two provide different types of protection. Chassis grounds are required for personnel safety; logic grounds provide a common reference point for the plant's electronic equipment.

The connection between chassis and logic grounds allows noise (voltage differentials) created by high-frequency surges to penetrate, via common mode, into the electronic equipment. Because modern PLC central processing units and associated auxiliary equipment operate with clock speeds in the 10- to 20-MHz range, even very fast transient surge voltages (in the microsecond range) can create problems.

In most cases, a plant with adequate surge protection and a low ground resistance will not experience problems. If a problem does exist for the plant PLC or other solid-state equipment, the solution is to either separate the chassis ground from the logic ground, or to tie these two grounds together with a very high (e.g., 100 megohms) impedance.[17] This will prevent the high-frequency lightning surge from damaging the electronic equipment. The high-impedance tie is preferable because it prevents the logic ground from accumulating a high voltage that might injure personnel.

Devices for Surge Control

Surge arrestors or supressors for protecting solid-state components are commercially available in several forms (see box, page 12-17). In addition, it should be emphasized that standard main-lead surge protection for hydro plants should consist of the following:

- Surge arrestor at the dead-end structure of the switchyard/substation and immediately in front of the power transformers;
- Surge arrestor on all medium-voltage main circuit breakers;
- Surge arrestor and capacitors on all medium-voltage motors;
- Isolation transformers for all variable-frequency controllers (VFCs); and
- Surge arrestor and capacitors for the generator.

Surge Suppressors for Solid-State Components

There are three types of surge suppressors for control equipment: gas tube, silicon-avalanche, and metal-oxide varistors.

Gas tubes divert the transient surge to ground, or, in the terminology of the trade, "crowbar" (flashover) at a voltage level ranging from 90 volts to several kV, depending on the particular model of gas tube. Although they have operational times in the nanosecond range, gas tubes are, in fact, slower than silicon-avalanche or metal-oxide varistor suppressors. Their advantage is that they can handle significantly larger surge currents than the other two suppressors.

Silicon-avalanche and metal-oxide varsitor suppressors are described as "clamping devices;" i.e., they "clamp" the surge transient voltage at a certain low level (e.g., 8V), and hold it there until its rated dissipation-energy level is exceeded (it burns out). Because clamping devices absorb the transient surge rather than diverting it to ground, they can accommodate less current than the gas-tube devices handle.

Silicon-avalanche and metal-oxide varistor suppressors are often integrated with gas-tube arrestors in "hybrid" surge protection devices. A hybrid surge protector combines in a single component the advantages of both suppressor types: the gas tube's capability of handling large surge currents, and the faster transient response of a silicon-avalanche or metal-oxide varistor.

Ratings and What They Mean

According to various manufacturers' literature, a gas-tube protector with a 90-V breakdown rating is rated to handle a 20,000-A current surge. This large current value may be deceiving. A gas-tube protector for control circuits is a very small device. The 20,000-A rating is for a one-time, 8/20 microsecond impulse wave form, which translates into an energy rating of 50 watt-seconds. In terms more familiar for power engineers, this is approximately 14 x 10^{-6} kilowatt-hours. To keep in perspective the ratings claimed by manufacturers for their surge protector devices, the ratings must be related to the microsecond wave-shapes from which they are derived. The configuration used to test surge protection devices can be either an 8/20 microsecond current wave form, 0.5 microsecond 100 kHz ring wave form, or a 1.25 MHz ring wave form.

When specifying surge protection devices for the instrumentation and control circuitry, the engineer should ensure that:

- The clamp voltage is below the peak operating voltage of the circuit;
- He understands the impulse wave form shape used for testing the surge protection device to arrive at its current rating; and
- Protection is provided for both normal and common-mode over-voltages.

Chapter 12

Figure 12-6 is a schematic showing the standard main-lead surge protection for a hydro plant. In addition to these standard provisions, extra protective measures are required for the PLC and its associated components for the conditions listed below.

Remote I/O Cards or I/O Racks External to the Powerhouse. All I/O racks located outside of the powerhouse require additional surge protection. During a thunderstorm, the probability is very high that lightning could cause an induced or capacitively-coupled surge in the connecting cables or shields. The greater the distance to the I/O rack, the more surge protection should be provided.

Because induced/capacitively-coupled surges travel in both directions, surge protection should be provided at both ends of the RS-232 or RS-422 (electrical characteristics defined by the IEEE) serial communication cable; i.e., at the cable entrance to the remote I/O rack and at the powerhouse PLC-cable entrance.

A typical example for hydro plants involves the headwater- and tailwater-level inputs, which are connected to the control room PLC I/O rack. The cables supplying these PLC inputs must have surge protection to avoid destroying the associated analog input card on the I/O rack.

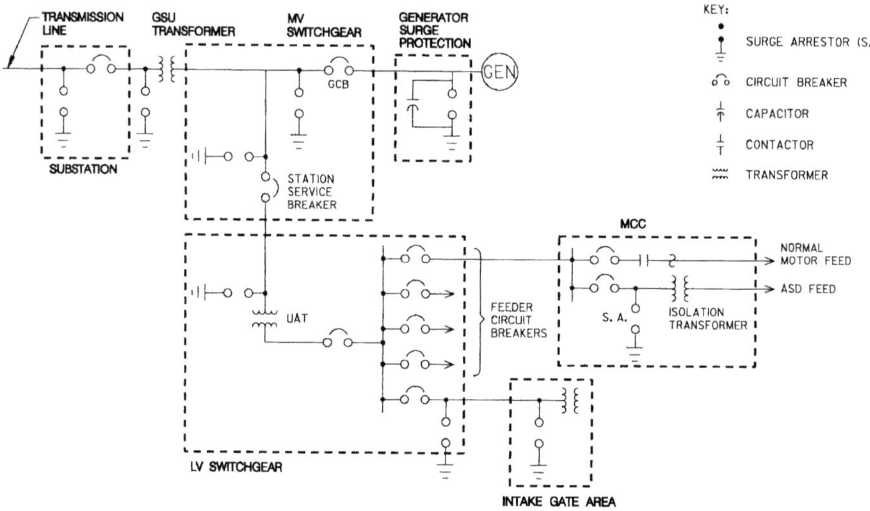

NOTES:

1. THIS LAYOUT ONLY ILLUSTRATES S. A. LOCATIONS FOR THE POWER SYSTEM. SURGE PROTECTION FOR THE CONTROL SYSTEM IS AN ENTIRELY SEPARATE ISSUE AND IS NOT ADDRESSED.
2. ALL MOTOR CONTROL CENTERS (MCC'S) THAT HAVE ADJUSTABLE SPEED DRIVES SHOULD HAVE AN ISOLATION TRANSFORMER AND SURGE PROTECTION FURNISHED AS AN INTEGRAL PART OF THE DRIVE EQUIPMENT.

FIGURE 12-6: ***Surge Arrestor Locations.*** *This schematic shows the typical location of main surge arrestors in a hydropower station.*

Telephone Lines/Modem Connections. Normally, the local telephone company will provide a "Positron Teleline Isolator" surge protector for the telephone system serving the plant. These devices were introduced to U.S. and Canadian telephone and power companies in 1978, and have been accepted as standard by many telephone utilities. If this equipment is in place, no additional protection is required for the phone system, including the PLC-modem connections.

If the telephone company does not provide this protection, the owner or engineer must determine what protection is furnished and decide what additional provisions are warranted.

Separate Ground Grids. An intake gate located a considerable distance from the powerhouse often is connected to its own ground mat system, which is separate from the powerhouse/switchyard system. In this case, any unbalance in the ground resistance between the two ground mats (intake gate and powerhouse/switchyard) will result in a differential voltage, just as a chassis-to-chassis differential voltage occurs between two chassis grounded at two different points.

If the distance from the intake-gate ground mat to the main ground mat is greater than one-half mile, running a ground cable to connect the two is not only expensive, but it frequently will not solve the voltage-differential problem. A voltage differential still will exist with the connection in place, although the differential will be lower in magnitude than if the mats were not connected. The reason is that the cable connection itself represents a certain impedance to any lightning-induced/capacitively-coupled current (see sidebar).

The solution is to ground the shield of the connecting RS-232 wire at both ends of the cable, in addition to providing surge protection at both ends of the cable.[16] Grounding the shield wire at both ends of a long cable run eliminates the induced high-frequency potentials usually associated with lightning strokes.[18]

An alternate solution is to use short-haul modems to convert low-level DC signals to AC, and then back to DC. The transformer coupling of the modems effectively isolates the signals.[17] A more recent solution is to use fiber-optic cables to provide the signal pathway from remote components to the powerhouse PLC.

Cable Shield Connections

Common practice over the years has been to ground the shield around the conductors at only one end; typically, the power source (feed) end. This procedure was a convenient way to obtain a single-point ground for the conductor shield, and it facilitated elimination of both 60-Hz voltage potentials and higher-frequency, capacitively-coupled voltage potentials.[19] Unfortunately, single-ended grounding will not protect the signal conductor from inductively-coupled voltage potentials, which are common mode, and can only be inhibited by grounding the conductor shield wire at both ends. Instances have been noted

Relationship Between Frequency and the Impedance of a Ground Conductor

The ground conductor impedance that occurs during a 60-Hz, line-to-ground short circuit is distinctly different from the impedance that occurs during a lightning surge to ground. At frequencies higher than 500 Hz, the reactive component of impedance in the ground conductor will increase significantly because of the inductance of the conductor.

Alternating-current systems are dependent on the ground impedance:

$$Z = [R^2 + (X_L - X_C)^2]^{1/2}$$

rather than just the resistive component of the impedance, which is the case for a direct current system. As frequency increases, the inductive reactance component ($X_L = \omega L$) dominates the equation and, in practical terms, determines the value of the system ground impedance. As impedance (Z) increases at higher frequencies, the voltage differentials between various electronic chassis assemblies increase. This means that although the voltage differential between two components may be acceptable at 60 Hz (during normal operation), the high frequencies associated with lightning strokes will create an intolerably high voltage differential.

To demonstrate the differences, this example compares a 500 A current for a 60-Hz line-to-ground short circuit to a 1-MHZ lightning-surge current. The ground conductor in both cases is a 10-foot length of 4/0 AWG bare copper cable.

Basic Equations:

$$E = IZ = I[R^2 + (X_L - X_C)^2]^{1/2}$$

$$X_L = \omega L$$

$$X_C = 1/\omega C$$

For a 4/0 AWG cable (Values extracted from Table 9 in *Grounding and Shielding Techniques in Instrumentation* [19]):

R= 49 millohms/1,000 ft

L= 540 microhenries/1,000 ft

C and X_C are neglected in both cases due to the high frequency.

For the line to ground fault:

$$E = 500A\,[(\,(49m\Omega/1{,}000\text{ ft})10\text{ ft})^2 + (6.28\,(60Hz)\,(540\text{ micro h}/1{,}000\text{ ft})10\text{ ft})^2]^{1/2}$$

$$= 500A\,[(.49\,(10^{-3}))^2 + (2.034\,(10^{-3}))^2]^{1/2}$$

$$= 1.02V$$

For the lightning surge to ground:

$$E = 500A\,[(\,(49m\Omega/1{,}000\text{ ft})10\text{ ft})^2 + (6.28(10^6Hz)(540\text{ micro h}/1{,}000\text{ ft})10\text{ ft})^2]^{1/2}$$

$$= 500A\,[(.49\,(10^{-3}))^2 + (33.9)^2]^{1/2}$$

$$= 16{,}950V$$

where grounding the shield wire at both ends has reduced the inductively coupled potential from higher than 10 kV to less than 500 V.[20]

The most recent practice concerning shield grounds is directly related to the greater amount of solid-state component equipment now located in hydroelectric plants. Ideally, electrical designers now ground the shield at the location that will create the minimal amount of transient voltage. To do this, the engineer/electrician must look at the overall system design and decide whether he is protecting against high-frequency, inductively-coupled potentials, capacitively-coupled potentials, or 60-Hz potentials. Although this is not always an easy task, it is essential if the engineer is to eliminate undesired potentials in the signal conductors.

Chapter 12

References

[1] Uman, Martin A., "The Lightning Discharge," *International Geophysics Series,* Volume 39, ISBN 0-12-708350-2, Academic Press Inc., 1987, pages 8, 13, 22, and 23.

[2] Corliss, William R., *Lightning, Auroras, Nocturnal Lights, and Related Luminous Phenomena,* ISBN 0-915554-09-7, The Sourcebook Project, Glen Arm, Maryland, 1982.

[3] Uman, Martin A., *All About Lightning,* ISBN 0-486-25237-X, Dover Publication Inc., 1986, pages 43, 55, 93, and 146.

[4] "Lighting Detection Network Averts Damage and Speeds Restoration," IEEE *Computer Applications in Power,* April 1996.

[5] *Lightning Protection Code,* Standard 78, American National Standards Institute/National Fire Protection Association, 1989.

[6] Marshall, J.L., *Lightning Protection,* ISBN 0-471-57305-1, John Wiley & Sons, Inc., 1973, pages 88, 93, 166, and 170.

[7] Viemeister, Peter E., *The Lightning Book,* ISBN 0-262-72004-3, The MIT Press, 1972, pages 67, 92, and 233.

[8] Richardson, Robert A., "Lightning Protection for High-Tech Buildings," *EC&M,* August 1989.

[9] Wagner, C.F., G.D. McCann, and C.M. Lear, "Shielding of Substations," Institute of Electrical and Electronics Engineers (IEEE) T*ransactions*, February 1942, Volume 61, pages 96-100.

[10] Gehrig, Edward H. and Ralph S. Gens, "Experience Revises Substation Lightning Protection," IEEE *Transactions,* June 1957, Volume III, pages 149-155.

[11] "IEEE Guide for Direct Lightning Stroke Shielding of Substations," IEEE 998, IEEE, 1996.

[12] Hartwell, Frederic P., "Council Nixes NFPA 781," *EC&M,* October 1995.

[13] "Master Labelled Lightning Protection Systems," Underwriters Laboratories' document UL 96A, Eighth Edition, July, 1974.

[14] "Recommended Practice For Grounding Industrial and Commercial Power Systems," IEEE Standard 142 ("Green Book"), ANSI/IEEE, 1991.
[15] "Recommended Practice For Powering And Grounding Sensitive Electronic Equipment," IEEE Standard 1100 ("Emerald Book"), IEEE, 1992.
[16] "Station Earthing, Safety and Interference Aspects, Final Report of Task Force 36.04.01," *ELECTRA* Magazine, No. 71, July 1990, pages 51 and 63.
[17] Hapten, James E., "Watch That Wiring," *Instrumentation and Control Systems,* April 1988.
[18] "Practical Applications of ANSI/IEEE Standard 80," *IEEE Guide for Safety,* 1986, page 87.
[19] Morris, Ralph, *Grounding and Shielding Techniques in Instrumentation,* 3rd edition, ISBN 0-471-83805-5, John Wiley & Sons, Inc, New York, New York, 1986, pages 152 and 153.
[20] Anders, R., B. Lindelof, and J. Nirs, "Problems with Interference Voltage in Control Equipment for Power Stations and Substations," CIGRE 1984 Session, Paper 36-08.

BIBLIOGRAPHY

Electrical Transmission and Distribution Reference Book, 4th edition, chapters 16 and 17, Westinghouse Electric Corporation, 1964.

Freedman, David H., "Bolts from the Blue," *Discover*, December 1990, pages 51-56.

Mousa, A. and R.J. Wehling, "Survey of Industry Practices Regarding Shielding Substations Against Direct Lightning Strokes," IEEE Power Engineering Review, January 1993.

Thaper, B., D.N. March, and Z.A. Alaywan, "Parametric Analysis of Lightning Caused Flashovers on H.V. Wood-Pole Lines," presented at the American Power Conference, April 1988.

Williams, Earle R., "The Electrification of Thunderstorms," *Scientific American,* November 1988, pages 88-89.

STANDARDS:

"Guide for the Application of Metal-Oxide Surge Arresters for Alternating-Current Systems," ANSI/IEEE Standard C62.22, ANSI/IEEE, 1991.

"Guide on Electrostatic Discharge; ESD Withstand Capability Evaluation Methods (for Electronic Equipment Subassemblies)," IEEE Standard C62-38, IEEE, 1994

"Guideline on Electrical Power for Automatic Data Processing (ADP) Installations," Federal Information Processing Standards (FIPS) Publication 94, 1994.

"IEEE Guide for Installation of Electrical Equipment to Minimize Electrical Noise Inputs to Controllers from External Sources," ANSI/IEEE Standard 518, IEEE, 1982.

"IEEE Guide for Safety in AC Substation Grounding," ANSI/IEEE Standard 80, IEEE, 1986.

"IEEE Guide for Surge Voltages in Low Voltage AC Power Circuits," ANSI/IEEE Standard C62.41, ANSI/IEEE, 1991.

"IEEE Guide for the Application of Gapped Silicon-Carbide Surge Arrestors for Alternating Current Systems," ANSI/IEEE Standard C62.2, ANSI/IEEE, 1987.

"IEEE Guide for the Application of Gas-Tube Arrestor Low-Voltage Surge-Protective Devices," ANSI/IEEE Standard C62.42, ANSI/IEEE, 1992.

"IEEE Guide on Surge Testing for Equipment Connected to Low-Voltage AC Power Circuit," ANSI/IEEE Standard C62.45, ANSI/IEEE, 1992.

"IEEE Standard for Gapped Silicon Carbide Surge Arrestors for AC Power Circuits," ANSI/IEEE Standard C62.1, ANSI/IEEE, 1989.

"IEEE Standard Test Specifications for Avalanche Junction Semiconductor Surge-Protective Devices," IEEE Standard C62.35, IEEE, 1987.

"IEEE Standard Test Specifications for Gas-Tube Surge-Protective Devices," ANSI/IEEE Standard C62.31, ANSI/IEEE, 1987.

"IEEE Standard Test Specifications for Low-Voltage Air Gap Surge-Protective Devices," ANSI/IEEE Standard C62.32, ANSI/IEEE, 1981.

"IEEE Standard Test Specifications for Varistor Surge-Protective Devices," ANSI/IEEE Standard 62.33, ANSI/IEEE, 1982.

National Electric Code, National Fire Protection Association (NFPA) 70/ANSI Standard C2, NFPA/ANSI, 1986.

"Standard Draft Application Guide For Surge Protection of Electric Generating Plants," IEEE Standard C62.23, IEEE, 1995.

"Standard for Metal Oxide Surge Arrestors for AC Power Circuits," ANSI/IEEE Standard C62.11, ANSI/IEEE, 1993.

"Standard Test Methods for Surge Protectors Used in Low-Voltage Data, Communications, and Signaling Circuits," IEEE Standard C62-36, IEEE, 1994.

"Surge Withstand Capability Test for Protective Relays and Relay System," ANSI/IEEE Standard 37.90.1, ANSI/IEEE, 1989

Chapter 13
Batteries at Hydropower Plants

Battery systems are installed in every hydroelectric powerhouse, large or small, to provide emergency power. Battery systems are one of the most crucial electrical systems in a hydro plant, because the battery system provides power to critical controls, protective relays, and uninterruptible power systems associated with computers that control plant operations. Reviewing the various types of batteries and battery chargers, electrolyte maintenance, and installation procedures is helpful in ensuring a reliable, long-lived emergency power source for a hydro plant.

Battery Types

Table 13-1 lists commonly used battery types and their normal expected life, approximate number of full discharges, ampere-hour range, approximate cost range, frequency of use, and advantages and disadvantages. The battery types are divided into two groups: flooded liquid electrolyte and sealed maintenance-free.

Flooded Liquid Electrolyte Batteries

These are the standard type of stationary batteries in hydro plants. Each cell is in a transparent "jar" that allows the plant operator to visually inspect the cell plates and electrolyte level. Low electrolyte levels can be restored by removing the vent cap and adding distilled water to the cell. When this type of battery is charging, it produces hydrogen gas.

These batteries, often referred to as lead-acid types, are reliable. Owing to the large ampere-hour capacity typically required in power plants, this type of battery historically has been used in hydroelectric facilities.

Sealed Maintenance-Free Batteries

Sealed maintenance-free batteries also are identified as "valve-regulated sealed lead acid" batteries. This type of battery was introduced in the 1970s for use in the telephone/communications industry. The individual cells are in flame-

Battery System Terminology

Equalizing Charge: An increased current output from the battery charger at a higher voltage level than the float charge voltage level at specified intervals of time to ensure that all cells remain in the fully charged condition.

Float Charge: The output from the battery charger at the float voltage selected to supply a continuous trickle charging current to the battery. This charging is set to maintain the cells in a fully charged condition. During float charging of the batteries, the battery charger also is supplying the normal direct current (DC) load requirements of the powerhouse. This is normally called "full float operation".

Flooded Liquid Electrolyte Battery: The traditional type of battery containing a liquid electrolyte in a transparent jar with a vent cap. (Commonly called a "wet" cell.)

Hydrometer: A transparent tube with a bulb on one end and an internal float. The bulb is compressed, the tube is inserted into the battery electrolyte, electrolyte is withdrawn by releasing the bulb, and the specific gravity is read from a calibrated scale on the side of the float.

Pilot Cell: One cell of a battery set that has the electrolyte specific gravity and temperature measured routinely every 30 days. This cell should reflect the condition of all other cells in the set. If the set consists of a number of cells on different battery rack elevations, one cell on each elevation should be designated as a pilot cell.

Sealed Maintenance-Free Battery: A battery that has cells containing either a gelled electrolyte (gel-cell) or a suspended electrolyte. The cells are equipped with a pressure relief vent that will open if a certain internal pressure is exceeded; the vent will reseal once the pressure is relieved. The batteries are not maintenance-free, although their maintenance requirements are less than the flooded liquid electrolyte batteries. These batteries are identified by the Institute of Electrical and Electronics Engineers as "valve-regulated sealed type" to avoid the misconception that they are maintenance-free over an entire life cycle.

Specific Gravity: The measure of weight of 1 liter of electrolyte compared to the weight of 1 liter of water at a specified temperature. In the specific case of batteries, this temperature is 25 degrees Celsius (77 degrees Fahrenheit).

retardant, shock absorbent, polypropylene plastic containers. Consequently, neither the cell plates or electrolyte level can be seen.

There are two specific types of sealed maintenance-free batteries: gelled electrolyte (also called the gel-cell), and absorbed (or starved) electrolyte. The gelled electrolyte type uses an electrolyte additive to immobilize most of the electrolyte between the plates in a semisolid gel. The absorbed electrolyte type uses a highly absorbent mat — either glass or polymer fiber — that immobilizes the electrolyte between the plates. Both types of sealed maintenance-free batteries have strengths and weaknesses.

TABLE 13-1: ***Commonly Used Battery Types in Hydroelectric Plants [2]***

Type	Normal Expected Life* (Years)	Approximate Number of Full Discharges	Ampere-Hour Capacity Range	Relative Cost of Battery	Advantages	Disadvantages
Flooded Liquid Electrolyte						
Lead Acid Antimony	15-20	1,000-1,200	6-4,000	Base	Capable of providing a significant number of full discharges over 20-year life	Frequent water addition, high hydrogen emission, needs monthly equalizing charge
Lead Acid Calcium**	20-30	50-100	6-4,000	Base	Low water consumption, low hydrogen emission, no monthly equalizing charge	
Lead Acid Plante	25-30	1,000-1,200	6-4,000	2.5x base	Can function at room temperature higher than the standard 77°F	
Nickel Cadmium Alkaline (NI-CAD)	20-25	1,200	2.5-1,000	3x base	Performs well in low temperatures, not damaged by freezing, low self-discharge rate, will not deteriorate in discharged condition, no release of corrosive fumes	At 1.2 volts per cell, requires a greater quantity of cells to attain a rating of 125-volt DC
Nickel Iron Alkaline	10-22	5,000-10,000	2.5-500	3x base	Performs well in low temperatures, not damaged by freezing, low self-discharge rate, will not deteriorate in discharged condition, no release of corrosive fumes	Not used in powerhouses; most common use is motive power application (golf carts, etc.)
Sealed Maintenance-Free						
Lead Acid Calcium	5	200	200-4,000	1.2x base	Does not require water addition, no hydrogen emission	Little historic operating experience in powerhouses
Lead Acid/ Special Alloy	20	300	200-4,000	1.4x base	Does not require water addition, no hydrogen emission	Little historic operating experience in powerhouses

* Lifetime estimates can vary substantially depending on cell/plate construction.

**This type of battery is the one most commonly used in hydroelectric powerhouses.

Chapter 13

The "maintenance-free" terminology is a misnomer, as these batteries must be inspected at regular intervals. However, compared to the flooded liquid electrolyte battery, maintenance requirements for the sealed type are minimal.[1] These batteries do not emit hydrogen gas during normal charging; however, the cell itself is equipped with a pressure relief vent that operates if a certain internal pressure is exceeded (to prevent the cell from exploding). This vent reseals once the pressure is relieved.

Vendor brochures often state that this type of battery has a 20-year life at "full float" and 25 degrees Celsius (77 degrees Fahrenheit). Other technical literature gives typical "expected life" values of five to 15 years. The total life span is related to cell/plate construction, number of full discharges, and maintenance of the battery. Owing to the lack of long-term historical operating experience data for the sealed maintenance-free batteries in powerhouses, prospective buyers should investigate each of these factors before purchasing batteries.

Selecting and Installing Battery systems

Sizing the Battery System

The size of battery system required in a powerhouse depnds on the ampere-hour capacity required to provide the needed emergency functions. An excellent example of an ampere-hour sizing calculation is given in the *Handbook of Electric Power Calculations*.[3] Another example is illustrated in the Institute of Electrical and Electronics Engineers (IEEE) Standard 485.[4]

Considerations for Physical Location

A common misconception among engineers is that a battery room, physically separate from other rooms in the powerhouse, must be provided for a battery set regardless of its ampere-hour size or the overall physical size of the powerhouse.

According to the National Electrical Code (NEC), batteries are to be located in an area that provides adequate ventilation and protection to personnel. In addition, any electrical equipment operating at 50 volts or more must be located in a room, vault, or similar inclosure that is accessible only to qualified personnel. Furthermore, batteries must be in a clean, dry room arranged to provide sufficient work space for inspection and maintenance.[5] However, the NEC does not explicitly require that an individual battery room, physically separate from the control room, be provided if adequate ventilation is available and only qualified personnel have access to the area. Moreover, the American National Standards Institute/Institute of Electrical and Electronics Engineers (ANSI/IEEE) standards do not require that a physically separate battery room be installed in each powerhouse.[4,6,7] Therefore, an engineer/owner must make the decision about the need for a physically separate, ventilated battery room on a site-specific basis.

In plants larger than 5 MVA, a ventilated battery room often is required. For these plants, a flooded liquid electrolyte type battery is normally selected. As a result, a large amount of hydrogen gas is generated during charging. The large ampere-hour size of this type of battery increases the amount of hydrogen gas during charging and contributes to the need for ventilation.

If a sealed maintenance-free battery is selected, a physically separate, ventilated battery room is not necessary because the battery produces no hydrogen. However, for most plants larger than 5 MVA, a separate battery room may be needed to accommodate the large physical size of the batteries and to ensure sufficient work space for inspection and maintenance.

At plants smaller than 5 MVA, a ventilated battery room is not always required for a flooded liquid electrolyte battery, because the small ampere-hour size means that the battery produces a minimal amount of hydrogen gas during charging. Furthermore, typically only qualified personnel (usually the operators) have access to the powerhouse. Therefore, according to the NEC, a separate battery room is not required.

Owners of plants smaller than 5 MVA should perform a "hydrogen produced" calculation to verify whether a ventilated battery room is necessary. Elements of this calculation are:

- Air volume in the powerhouse;
- Number of air changes per hour based upon the heating, ventilating, and air conditioning (HVAC) system design; and
- Amount of hydrogen gas released during charging. (This amount is based upon the ampere-hour size of the battery, the float charging voltage selected, and the type of battery to be charged.)

This calculation should confirm that, in the worse-case scenario, the hydrogen production of the batteries will be less than 1 percent of the total air volume in the powerhouse. (This criterion conforms with Occupational Safety and Health Administration requirements.) Normally, a minimum of 3 percent hydrogen by volume is required before any explosive mixture is possible.

Temperature Considerations

If a battery room is installed, the HVAC system should maintain the temperature in the room at approximately 25 degrees Celsius (77 degrees Fahrenheit) to ensure that the rated output of the battery is obtained. Higher temperatures increase the rate of chemical reactions within the battery and shorten its life. Lower temperatures reduce the rated output of the battery (similar to a car battery in cold weather), and can adversely affect the ability of the power plant to return to service during a worse-case scenario.

If a battery room is not required due to the low level of hydrogen gas emitted, plant managers and operators should be aware that reducing the powerhouse ambient temperature in the wintertime to conserve heat will adversely affect the battery. This problem occurs frequently at unattended small hydro installations.

FIGURE 13-1: ***Batteries in the Powerhouse.*** *Lead-acid type batteries, such as those shown in this photograph, historically have been used in hydro facilities as emergency back-up systems. These batteries are considered very reliable, which makes them a popular battery choice in power plants. Also known as flooded liquid electrolyte batteries, lead-acid batteries produce hydrogen gas during the charging mode, so battery rooms must be adequately ventilated. The NEC/ANSI standards do not require the equipment in the room to be explosion-proof.*

Electrolyte Maintenance of Batteries

To deliver rated output, a battery electrolyte consists of a nominal 30 percent sulfuric acid and 70 percent distilled water solution at a specific gravity of 1.215 and an electrolyte temperature of 25 degrees Celsius (77 degrees Fahrenheit). If the electrolyte temperature varies from 25 degrees Celsius by more than plus or minus 1 degree, the specific gravity readings must be multiplied by a correction factor to reflect the actual specific gravity of the electrolyte. These correction factors usually are noted on the battery manufacturer's standard historical record sheets.

When the battery supplies current to a load, some of the sulfuric acid of the electrolyte is consumed and creates lead sulfate on the negative and positive plates of the cell. Therefore, the percentage of sulfuric acid in the electrolyte decreases. This decrease in sulfuric acid content is reflected by a corresponding decrease in the electrolyte specific gravity. To determine the state of charge that exists in a battery set, engineers often use a hydrometer to measure the specific gravity of a designated pilot cell.

When the battery is being recharged by the battery charger(s), some of the water portion in the electrolyte is converted to hydrogen and oxygen gases (known as "bubbling" of the electrolyte). Consequently, this charging results in a lower electrolyte level in the cells and a higher specific gravity. The electrolyte now has a percentage of sulfuric acid higher then the nominal 30 percent in a fully charged battery cell. Therefore, additional distilled water must be added to the cell to return the electrolyte to the nominal specific gravity of 1.215.

Typically, distilled water is added every six to 12 months to a lead acid calcium type battery maintained under a "float" charge. If the load discharges related to emergency situations occur and the "equalizing" charge mode of the battery charger is used to rapidly restore the batteries to their nominal condition, the battery cells will require additional distilled water.

Battery Chargers

In addition to charging the batteries, battery chargers supply a plant's normal direct current (DC) and essential alternating current (AC) loads through an inverter. Although ANSI/IEEE Guide 1020, "Guide for Small Hydroelectric Power Plants," states that a redundant charger is not required in small plants, a second charger can prevent a single point failure (one battery charger) from jeopardizing the reliability of the power plant's ultimate emergency power source (the station batteries).[8] Where practical, a spare, transportable unit can be used more economically as a backup for several plants rather than purchasing redundant battery chargers for each plant.

Historically, a number of methods have been used for charging stationary batteries. The primary methods used today are either constant-potential or constant-current.

The constant-potential method monitors the voltage of the battery and automatically tapers off the amount of charging current as the cell approaches the fully charged condition. This method minimizes hydrogen gas evolution, and is the one most often used at hydroelectric facilities.

The constant-current method will feed a constant charging current into stationary batteries for a certain period of time regardless of the cell voltage. This charging method is prevalent in industrial applications where batteries are continuously discharged.

Two Rates of Charging

A constant-potential charger should have two rates of charging: the "float" charge and the "equalizing" charge. The float charge is a continuous trickle charging current applied to the stationary battery. The amount of required float current varies, depending upon the battery potential sensed by the battery charger. The various types of stationary batteries will have different "self discharge" rates depending upon their construction, age, and the electrolyte specific gravity. The constant-potential type battery charger will adjust the amount of trickle current for these various conditions when in the float charging mode. Therefore, this mode assures that the battery will be at full capacity when an emergency condition arises.

Periodically, the stationary battery must be given an equalizing charge at a higher voltage than the float voltage to ensure that all cells are fully charged. This is because individual cells, over a period of time, tend to differ in their ability to uniformly accept a float charge, probably due to small, unavoidable chemical or physical differences within the cells. When this situation occurs, some cells will become fully charged before others do. Consequently, the charger's constant potential sensing circuitry will decrease the amount of trickle charging current before the entire battery is fully charged.

The equalizing charge current is at a higher voltage and larger current than float charging. It provides a mixing action within the electrolyte of each cell, which ensures electrolyte uniformity. This minimizes the individual cell differences, and will make the float mode of the charger more effective. This equalizing charge also is used to rapidly recharge the battery after discharging during an emergency.

Voltage Regulation

The charger should be capable of supplying a float or equalizing charge with no more than a plus or minus 1 percent voltage variation throughout its ampere rating (i.e., from no-load to full-load output). This plus or minus 1 percent voltage variation is to be maintained with a plus or minus 10 percent supply voltage variation and over a plus or minus 5 percent supply frequency variation. Precise voltage regulation is required because minor voltage variations on the charger output will drastically change the amount of charging current sent to the batteries, resulting in either battery overcharging or undercharging. The close monitoring (plus or minus 1 percent) of the output voltage indicates that the charger is *regulated*, and that it will maintain a constant output voltage regardless of the line voltage or DC load current variations.

Battery chargers for sealed maintenance-free batteries also must include temperature compensation of the charger float voltage. Attaining the exact float voltage is more crucial for sealed cells than for flooded cells.[1] The sealed cells have only a certain amount of internal material for use during the charging process. Once this material is used, continuing to charge the cell will result in decreasing the battery's duty life. Therefore, it is imperative that the charger float voltage setting be even more closely regulated for sealed, maintenance-free batteries than for the standard flooded liquid electrolyte type batteries.

Automatic Switching from Float to Equalizing Mode

Switching from the float to equalizing mode should be done automatically. However, chargers can be purchased with only manual mode switching. In manual mode, an operator switches the charger from the float mode to the equalize mode once the battery has been discharged and back to float mode once the battery is fully charged. However, this reliance on an operator's availability and/or discretion will, over a period of time, result in charging errors that shorten the battery's life. If the errors are serious, such as leaving the charger in the equalizing mode for an extended period and consequently "bubbling off" a majority of the electrolyte, the battery's recharging capability will be seriously impaired. This will have the effect of jeopardizing plant availability and energy production because the battery will not be capable of producing at rated capacity.

Automatic switching of the charger (commonly called automatic mode switching) uses voltage sensing to automatically switch the charger to the equalizing mode after the battery has been discharged. Once the battery cell potentials rise to approximately 90 to 95 percent of their rated potential, the charger automatically switches to the float mode. Consequently, the equalizing charge is only applied to the batteries for the amount of time required to obtain 90 to 95 percent of the rated cell potential. This approach minimizes hydrogen gas evolution and prolongs battery life.

Charger Sizing and Options

To determine the current capacity of the battery charger, the following formula can be used:

$$\text{Charger DC amperes} = \text{constant ampere load on system} + (1.1)C/N$$

where C is the calculated number of ampere-hours discharged from the battery, and N is the number of hours to recharge the battery.

A battery charger for powerhouse operation should have several features. The following options are the basics:

- Automatic two-rate, float-equalize selection with voltage sensing (automatic mode switching);
- Plus or minus 1 percent voltage regulation from no-load to full load with a plus or minus 10 percent supply voltage variation and plus or minus 5 percent supply frequency variation;
- Temperature compensation for the float charging circuit, if the charger is to be used with sealed maintenance-free batteries;
- AC power loss relay and AC ground overcurrent relay;
- High/low DC voltage output relays; and
- Positive/negative ground detection relays.

Different battery types have different recommended float charge ranges as identified by the manufacturers. For powerhouses, it is desirable to select a float

Chapter 13

charging value at the high end of the specified range to eliminate the need for frequent equalizing charges.

Avoiding Common Problems

Installation Procedures

The most common problem with batteries is poor physical installation procedures that lead to battery failure months or years later. Installation problems that frequently occur include:

- Improperly torqued bolts on the battery connections;
- Damaged terminal posts or post seals due to incorrect handling of the batteries;
- Unlevel battery rack;
- Inadequate space for cleaning and inspecting the cells; and
- Incorrect ground connections.

A good installation procedure that will avoid most of these problems is detailed in IEEE Standard 484.[6] The contractor installing the battery should be intimately familiar with this procedure.

In general, poor physical installation for battery chargers should not pose a problem, as the chargers are solid-state equipment components that are usually wall-mounted in the immediate area of the batteries. Installation problems with chargers are usually related to improper float and equalizing charge set points and improperly set protective relays.

Battery Monitoring Systems

Battery maintenance consists of periodic maintenance inspections every 30 days to confirm the following:

- Electrolyte level;
- Pilot cell specific gravity;
- Pilot cell voltage; and
- Tightness of intercell connections

In addition, *annual* maintenance inspections require that the specific gravity and voltage of each cell be checked, and that the resistance of the intercell connections be measured.

The maintenance inspections detailed above are labor-intensive. Consequently, a battery monitoring system (BMS) was developed in 1994 to automatically monitor the health of each battery cell. BMS equipment automatically measures each cell's voltage, temperature, specific gravity, internal cell resistance, external intercell resistance, and state-of-charge. In addition, during an equalizing charge, the BMS equipment will automatically divert charging current around a cell that is already fully charged (2.25 volts per cell for lead-calcium batteries). This diversion of current avoids "bubbling" or overcharging of the fully charged cells in a 60-cell battery set.

The BMS equipment is externally attached in a sealed module to the individual battery cell. The BMS provides "on-line" condition monitoring of the health of each cell using a communication link connected to a programmable logic controller or personal computer. Typical BMS packages include software to query and store the information.

The BMS equipment is currently available on the market; however, at present it is not common at hydro plants. Most BMS units installed have been at unmanned telecommunication centers and substations.

Battery Testing Procedures

Battery testing is defined in IEEE Standard 450, which lists the following capacity tests for a battery system:[9]

Acceptance Test. This is a constant-current capacity test performed at the factory prior to shipment or after initial installation. Its purpose is to confirm the battery's ampere-hour capacity.

Performance Test. This is a constant-current capacity test performed at the site. IEEE 450 recommends a performance test every 10 years after the initial installation performance test has been completed to confirm the battery capacity.

Service Test. This test is a variable-current capacity test that simulates the actual time current profile (duty cycle) the battery was designed to supply, in accordance with IEEE 450. A service test should be performed only if a battery is suspected of being at the end of its service life (80% of rated capacity).

Alternative Test Method

For installed batteries, a performance test or service test as designated in IEEE 450 is very expensive. The tests utilize complex load banks and bulky measuring equipment. In addition to the expense of the tests ($20,000 - $40,000 per test), they require the battery set to be out of service for 8 to 16 hours.

An alternative method, introduced in 1995, lessens the demand on time and equipment by testing only one or two cells of the battery set. The new method reduces the frequency of "performance" and "service" tests required on the complete battery set. This method, however, has not been accepted by IEEE for inclusion in Standard 450 (as of 1998).

The entire test can be performed using lightweight portable equipment, and does not disable the entire battery set. The test begins by measuring the impedance of all the battery cells. The cell impedances are compared to previous impedance readings to check for increases in impedance. If cell impedances are the same as in previous test measurements, the battery's capacity has not changed and additional capacity tests are not required. The cells with the highest impedance (lowest capacity) and lowest impedance (highest capacity) are then selected from the

Chapter 13

battery set and discharged utilizing a current capacity test. If the measured capacity test is greater than 90% for both cells, then the battery set capacity is assumed to be greater than 90%. If the capacity test is less than 90% for either or both cells, the more expensive "performance test" should be performed on the entire battery set.[10]

References

[1] Migliaro, Marco W., "Application of Valve-Regulated Sealed Lead-Acid Batteries in Generating Stations and Substations," American Power Conference, April 1988.

[2] Higgins, Lindley R., *Maintenance Engineering Handbook,* McGraw-Hill Book Co., 4th Edition, Chapter 7, 1988.

[3] Hicks, T.G., A. Seidman, and H. Mahrous, *Handbook of Electric Power Calculations,* McGraw-Hill Book Co., 1983, Chapter 18.

[4] "Recommended Practice for Sizing Large Lead Storage Batteries for Generating Stations and Substations," Institute of Electrical and Electronics Engineers (IEEE) Standard 485, IEEE, 1983.

[5] National Electrical Code Handbook, Sections 480 and 110-17, 1996.

[6] "Recommended Practice for Installation Design and Installation of Large Lead Batteries for Generating Stations and Substations," IEEE Standard 484, IEEE, 1987.

[7] "Recommended Practice for Emergency and Standby Power Systems for Industrial and Commercial Operations," American National Standards Institute(ANSI)/IEEE Standard 446, ANSI/IEEE, 1995.

[8] "Guide for Small Hydroelectric Power Plants," ANSI/IEEE Guide 1020, 1988.

[9] "Recommended Practice for Maintenance, Testing, and Replacement of Vented Lead-Acid Batteries for Stationary Applications," IEEE Standard 450, IEEE, 1995.

[10] Gabriel, Charles M. and Uhlir, Kurt. W., "Simplifying Battery Testing," *Transmission and Distribution,* September, 1995.

Bibliography

Berndt, D., "Maintenance Free Batteries," ISBN No. 0-47-1-93960-9, John Wiley & Sons, 1993.

"Guide for Selection of Valve-Regulated Lead-Acid (VRLA) Batteries for Stationary Applications," IEEE Standard 1189, IEEE, 1996.

IEEE Sourcebook on Lead-Acid Batteries and Stationary Battery Standards Collection Set, IEEE, 1993.

Linden, David, "Handbook of Batteries," 2nd edition, ISBN No. 0-07-037921-1, McGraw-Hill Book Co., 1995.

"Recommended Practice for Maintenance, Testing, and Replacement of Nickel-Cadmium Storage Batteries for Generating Stations and Substations," IEEE Standard 1106, IEEE, 1995

"Recommended Practice for Installation Design and Installation of Valve-Regulated-Lead-Acid Storage Batteries for Stationary Applications," IEEE Standard 1187, IEEE, 1996.

"Recommended Practice for Maintenance, Testing, and Replacement of Valve-Regulated Lead-Acid (VRLA) Batteries for Stationary Applications," IEEE Standard 1188, IEEE,1996.

Chapter 14
Uninterruptible Power Supply Systems

Hydroelectric plants use uninterruptible power supply systems to provide electricity for essential loads when normal plant power systems fail. Loss of power to such loads as the plant computer, communications networks, security systems, and emergency lights can interfere with plant control, endanger personnel, and compromise operational and emergency responses.[1]

Hydro plant owners and engineers have a wide variety of choices for installing or upgrading uninterruptible power supply (UPS) systems, depending on the demands of the particular plant, the budget, and other planning variables. However, because there is no comprehensive standard for such systems, the wide variety of choices can create confusion among project owners and engineers. In that circumstance, the project owner or operator's best defense is a careful evaluation of plant needs and precisely written specifications.

There is no comprehensive ANSI/IEEE UPS standard that covers the primary UPS technologies and their hybrids. ANSI/IEEE 944, which was reaffirmed in 1996, defines the applications, performance requirements, output requirements, and testing requirements for a "static" UPS type system.[1] ANSI/IEEE 446, the "Orange Book," is the only standard that addresses "rotary" UPS systems.[2] Although this standard defines the different types of rotary UPS, it does not define performance requirements, output requirements, or testing requirements for a rotary UPS.

Similarly, there is no comprehensive international standard for uninterruptible power systems. This means that terms defined by the ANSI/IEEE standards and commonly utilized in the U. S. to define UPS equipment may not have the same meaning to international equipment suppliers.

Types of UPS Systems

The American National Standards Institute (ANSI) and Institute of Electronics and Electrical Engineers (IEEE) define a UPS system as one designed to automatically provide power, without delay or transients, during any period when the normal power supply is incapable of performing acceptably.[2]

Glossary of Power Supply Terms

Bypass Transformer: A bypass transformer provides alternating current (AC) power to uninterruptible power source (UPS) loads when the UPS equipment fails, is temporarily overloaded, or is out of service for maintenance.

Bypass Transformer K Factor: The K factor is a complicated term utilized to derate transformers to ensure the transformer does not overheat due to harmonics when supplying the solid state/non-linear loads associated with an uninterruptible power source (UPS). The K factor is used with fully rated transformers (kVa rating of the transformer is equal to or slightly greater than the load kVa), and becomes less useful as the percentage of the transformer kVa utilized decreases. The higher the K factor of the transformer, the higher the cost.

Crest Factor: The crest factor is the ratio of the crest voltage (peak maximum) to its root mean square (RMS) voltage (for a normal sine wave the crest factor is 1.414). The ratio is used to determine ruggedness of an uninterruptible power source (UPS) and its capability to furnish power to non-linear loads. A typical crest factor value for various manufacturers is 3.0. The higher the crest factor, the more capability the UPS has to supply non-linear loads.

Harmonic Distortion: Harmonic distortion is characterized by the appearance of harmonics in addition to the fundamental component. The total harmonic distortion is the ratio of the RMS value of all the harmonics to the RMS value of the fundamental.

Non-Linear Loads: Non-linear loads have such characteristics that with an applied sinusoidal voltage the load current is not sinusoidal. Non-linear loads are related to saturated magnetic circuits (transformers and motors) or solid-state devices (variable frequency drives or battery chargers) where the power consumption of the device is no longer directly proportional to the voltage applied due to saturation or harmonic production, and consequently causes a distortion in the voltage waveform.

Off-Line UPS: An off-line uninterruptible power supply (UPS) remains dormant until called upon to operate. An example is a static/rotary UPS system that uses a very fast static transfer switch. *On-Line UPS:* An on-line uninterruptible power supply (UPS) operates at all times. An example is a static UPS system supplied by 125-V direct current power from a rectifier/battery set arrangement.

Step Load Change: A step load change is an instantaneous change in load current (for example, zero to full load) for measuring the load regulation and recovery time of an uninterruptible power source.

Transient: A transient is a momentary voltage disturbance that affects the uninterruptible power source system's capability to deliver an acceptable sine wave output. Transients are defined in terms of their amplitude and time duration.

Uninterruptible Power Supply System: An uninterruptible power supply system converts unregulated input power to a controlled alternating current (AC) power voltage and frequency that is filtered to exclude extraneous noise and transients. The system continues to supply power without interruption when normal AC power fails.

Vital Loads: Vital loads are instrumentation and control power systems or other loads that are important to plant operation or personnel safety, or both.

In general, UPS systems are divided into static and rotary systems. The first such systems, which date from the late 1940s, were rotary motor-generator sets. Some of this equipment is still in use today in older hydroelectric stations. In the early 1960s, static UPS systems using thyristors (then known as silicon-controlled rectifiers) were introduced. These static systems dominated the UPS market from the 1960s through the early 1980s, when advances in technology made rotary systems competitive again.

Rotary Systems

Modern rotary UPS systems are available in a variety of configurations, among them motor-generator sets with flywheels for energy storage, motor-generator sets with static inverters, and motor-generator sets with induction coupling for momentary energy storage.[3] (Standard diesel generator sets often used in hydro plants serve as emergency standby systems, not UPS systems. Such systems typically provide electricity after a power outage of some 30 to 90 seconds duration, during which the diesel generator set starts, comes up to synchronous speed, and automatically connects to the electrical system.)

Rotary UPS systems typically include an alternating current (AC) motor-generator set with a flywheel, a rectifier, an inverter, storage batteries, a static transfer switch, a manual transfer bypass switch, and solid state control circuitry.

Static Systems

The most common static UPS systems use either pulse width modulation (PWM) or ferroresonant technology to achieve a stable 120-V AC output. PWM systems have more solid-state circuitry and generally require cleaner operating environments than the ferroresonant type. As a consequence, the ferroresonant type is considered more rugged. However, both types are capable of providing excellent UPS systems if the appropriate technical specification is written. Hybrid UPS systems combine PWM and ferroresonant technology.

In general, static systems consist of a rectifier, an inverter, storage batteries, a static transfer switch, a bypass input transformer, a manual transfer bypass switch, and solid-state control circuitry.

Comparing the Systems

There are advantages and disadvantages to each type of system. Rotary UPS systems usually are used when a large power capacity and exceptional noise isolation are required. Rotary UPS options typically cost 25 percent more than comparable static systems. Moreover, a rotary system's mean time between failures (MTBF) rate is lower than that of a static UPS.

Owners and engineers should remain aware that there is no ANSI/IEEE standard for rotary UPS systems equivalent to the standard for static UPS systems.

A June, 1988 article in the magazine *EC&M* provides additional information on the advantages and disadvantages of each type of system.[4]

Choosing a System Configuration

Both static and rotary UPS systems can be categorized into off-line and on-line systems. As defined by ANSI/IEEE, an off-line system remains dormant until called upon to operate, while an on-line system operates at all times.[5]

Off-Line Systems

Off-line systems also commonly are known as standby UPS systems. Off-line systems further are divided into "forward-transfer" and "continuous" systems. Both kinds of off-line systems rely on similar features to give them the ability to come on line within a few operating Hertz: static transfer switches that operate in less than 0.25 Hertz and very responsive solid-state sensing of the normal plant 120-V AC power source. Unfortunately, this very short transfer period is beyond the capability of most plant computers, and thus limits the usability of off-line UPS systems in many hydropower plants.

Forward-Transfer Systems. A forward-transfer system uses normal plant 120-V AC to feed the essential loads until the system circuitry senses a power failure. The system then switches to a battery-powered backup system. No power conditioning of the UPS sine wave output is performed until the battery-powered backup system is engaged. Figure 14-1 illustrates a typical forward-transfer system.

Continuous Systems. A continuous off-line UPS feeds the normal plant 120-V AC to the output of the UPS and, through a power-conditioning module furnished as an integral part of the UPS, to the plant's essential 120-V AC loads.

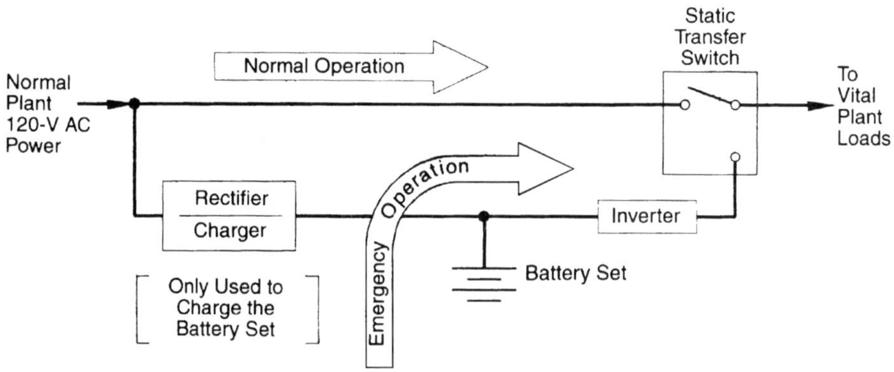

Figure 14-1: ***Forward-Transfer Off-Line UPS.*** *This drawing illustrates a typical forward-transfer, off-line system. Upon power failure, the static transfer switch opens at the normal plant AC (alternating current input) and closes into the inverter, thereby continuing to supply power.*

When the normal plant 120-V power fails, the system switches to a battery-powered backup system. Therefore, the continuous off-line UPS provides conditioning to the UPS sine wave output even when it is off line.

Using A Power Line Conditioner Instead Of An UPS

An UPS is normally required for power stations because the PC or PLC "ride-through" time (time the PC or PLC can operate without power) is less than 15 milliseconds. If the owner is confident that the power station is not prone to power outages or significant disruptions, a power line conditioner without battery back-up may be used. A power line conditioner simply takes the plant's normal 120-V AC and "conditions" the power by eliminating surges, noise, and voltage sags. However, the response time for a power line conditioner is in the range of 25 to 160 milliseconds, considerably more than the time required for an UPS to perform the same tasks.

A power line conditioner with battery back-up would include a static transfer switch and inverter. This arrangement would be classified as a "continuous off-line" type of UPS.

On-Line Systems

An on-line UPS system continually feeds power to the essential loads and power-conditions the UPS sine wave output to ensure a true sine wave. A static transfer switch is used to change to bypass power (normal plant 120-V AC) only when a problem develops in the inverter, rectifier, or other UPS component. Static on-line UPS systems often are referred to as "reverse-transfer" systems in the power industry. These are the most commonly used systems in hydro plants.

One common example is a static inverter system supplied by a 125-V DC battery bank. Figure 14-2 shows the configuration of a typical reverse-transfer on-line UPS system.

Off-Line Systems Versus On-Line Systems

A hydropower project owner or engineer should base the choice between an off- or on-line system on an analysis of the period of time essential plant systems can be without power (availability) and the number of power failures the essential plant system can tolerate (reliability). On-line UPS systems normally are used for essential 120-V AC systems in hydro plants that cannot tolerate any interruption of power — in other words, where 100 percent availability and reliability are required. Where the 120-V AC system can tolerate a few cycles of interruptible power (100 percent reliability; available 99-plus percent of the time based upon static transfer switch operating time and synchronizing speed), an off-line UPS system with a very fast switch normally can be used. A more detailed discussion of emergency power system availability and reliability is included in the ANSI/IEEE Gold Book.[5]

Chapter 14

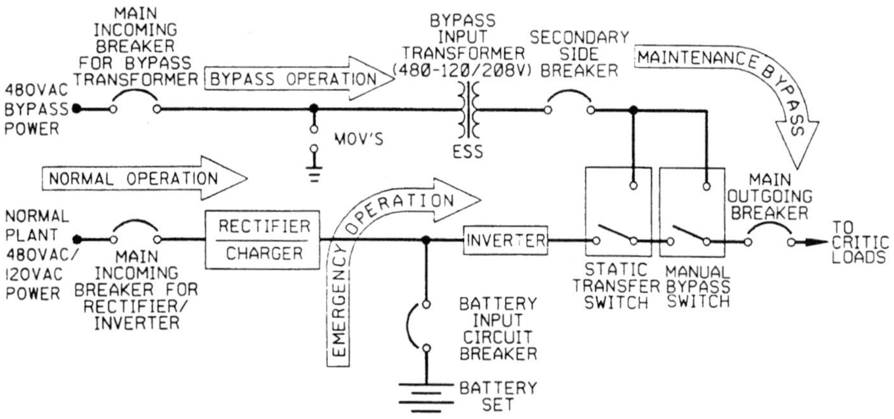

FIGURE 14-2: ***On-Line "Reverse-Transfer" UPS.*** *The static on-line UPS continually feeds electricity to computers, safety systems, and other essential loads. Power flows are shown for the following conditions:*

- *Normal operation*
- *Emergency operation due to loss of normal plant 120-V AC power*
- *Bypass operation due to failure of the rectifier or inverter*
- *Maintenance bypass operation*

The electrostatic shield (ESS) and surge arrestors (metal oxide varistors or MOVs) for the bypass input transformer are also shown.

North American hydro plant owners and operators should be mindful that, although ANSI/IEEE has defined the essential elements of off- and on-line UPS systems, there is no international definition of these terms. In fact, no internationally accepted standard for UPS systems exists. As a result, the purchaser should ensure that any technical specification for a UPS system requires all bidders to supply the same type of system, and that the specified system provides the reliability and availability necessary for the load it is to supply.

Power Output

ANSI/IEEE standards define the power output tolerances for a UPS.[2] The standard lists the typically accepted range of inputs that computers/computer equipment will sustain without malfunctioning. It also dictates the UPS input and output tolerances for various conditions.

SIZING HYDRO PLANT UPS SYSTEMS

Although ANSI/IEEE standards identify a number of vital power plant instrumentation and control categories, many are not applicable to hydropower plants.[6]

The size of an UPS system for a specific project depends upon a number of

TABLE 14-1: ***Normal Range of Sizes for Hydro Plant Uninterruptible Power Supplies***

Plant Size (MW)	UPS Size (kVA)
Less than 5	1 or less
5-25	1-5
25-100	5-10
100 plus	10 plus

variables, among them plant size and number of units. For plants larger than 100 MW, the kVA size of the UPS is highly dependent on the type of computer system in use and the physical layout of the powerhouse. The typical range of sizes for hydro plant UPS systems is shown in Table 14-1.

To size a plant UPS, project owners or engineers should tabulate all plant equipment to be fed from the system. The tabulation should identify each item of equipment and its volt-ampere (Va) load. Once the equipment has been tabulated, the total Va required for the plant can be determined by adding all the equipment Va loads. Once that has been determined, a 25 to 35 percent load growth factor should be added to the total Va for new plants to allow for future load growth. For existing powerhouses where UPS systems are being replaced, a 10 percent load growth factor should be more than adequate, unless a significant plant upgrade is scheduled in the future.

Once the kVA size of the UPS system is determined, the owner/engineer must determine the associated size of the battery set that will supply 125-V DC power to the UPS in an emergency. Emergency time periods of two to four hours are normally used at hydroelectric plants. The battery should be sized to supply the UPS kVA load for the period selected (see Chapter 13).

Many UPS systems with capacities of less than 1 kVA are sold as off-the-shelf packages. If such equipment meets the project needs based on the size evaluation, an owner can buy one of these standard UPS systems without writing detailed specifications.

WRITING SPECIFICATIONS FOR A UPS

Specifying Components

The technical specification for a hydro plant UPS system should include a detailed list of all items to be furnished. By including a complete list, the project owner can ensure delivery of a completely functional UPS system that can be integrated easily into the plant 125-V DC, 480-V AC, and 120-V AC systems. As a minimum, the following equipment components are critical for a static on-line reverse transfer UPS.

Rectifier/Inverter. The rectifier is the solid-state device that converts AC power into DC power to supply the input power to the inverter. The inverter is the solid-state device that changes the DC power received from the rectifier into AC output power (usually single-phase 240/120-V AC or three-phase 208-V AC). Both devices use thyristors to accomplish their function.

Manual Bypass Transfer Switch. The manual bypass switch, also commonly called the maintenance bypass switch, is used to bypass the entire UPS during maintenance periods.

Static Transfer Switch. For an on-line UPS, the static transfer switch transfers the UPS from its normal input to 120-V AC plant power upon failure of the UPS. For off-line UPS, the switch transfers from the normal 120-V AC plant power to the battery system on plant power failure. An auto-retransfer capability should be included to allow the switch to reconnect to its normal source once normal power has been restored.

Batteries. There are numerous types of flooded cell or sealed, maintenance-free batteries available, depending on space, physical location of the UPS system, and discharge capability. Selection of sealed maintenance-free batteries must be closely coordinated with the choice of charger to avoid over- or undercharging the batteries. In addition, if sealed maintenance-free batteries are mounted inside enclosures, the possibility of a battery thermal runaway should be thoroughly investigated.

Bypass Input Transformer. Normal 480/120-V AC plant power is connected to an on-line UPS through a bypass input transformer to supply power during an UPS inverter/rectifier malfunction. This transformer must be constructed to provide both transient voltage surge suppression and voltage stability. Engineer/owners frequently neglect the specification for the transformer in preparing UPS system specifications. The specification should require a regulating transformer that has the following characteristics:

- Noise immunity: electrostatic copper shield should be provided between the primary and secondary windings to shunt common mode noise to ground;
- Transient protection: surge suppression devices, usually metal oxide varistors, are required on the primary pole of the transformer to prevent high-energy spikes from penetrating through the transformer;
- Voltage stability: the voltage regulation of the transformer should be less than 1 percent to avoid any significant change in the input/output voltage ratio at lower power factors;
- kVA Rating:the transformer should be adequately rated in relation to the UPS kVA output rating to ensure it does not have temperature-related problems at full load and will not add any distortion to the output voltage waveshape; and
- K Factor: transformers closely sized to the total load requirement should be provided with a K factor rating, a term used to derate a transformer to ensure it does not overheat due to harmonics when supplying the solid state/non-linear loads associated with an UPS.

Circuit Breakers. Several circuit breakers should be specified for safe operation: a main incoming breaker to the static rectifier/inverter; a main outgoing breaker from the static rectifier/inverter; a main incoming breaker for the bypass input transformer; a bypass source secondary side circuit breaker to allow the static transfer switch to be isolated; and a battery input circuit breaker to allow isolation from the battery source.

Local Instrument, Remote Alarm Indicators. UPS system manufacturers offer a wide variety of optional instruments/indicators and remote alarm contacts. These optional devices should be selected as required for the existing power plant.

Specifying Factory Tests

The technical specifications also should explicitly define required factory tests to ensure that the product specified is the one received. Routine and design tests for static UPS systems are defined in ANSI/IEEE standards.[1] The standards do not identify routine or design tests for rotary UPS systems; thus, such tests must be detailed in the engineers technical specification. It is essential that the owners engineer or other representative witness the factory tests. Table 14-2 lists the minimum factory tests that should be required.

The technical specification should also include a paragraph ensuring the UPS output kVA can be supplied over a power factor (PF) range of 0.8 to 1.0. Many manufacturers quote rated kVA at a 0.8 PF only. Including the 0.8 to 1.0 PF requirement dictates that a 15-kVA UPS will supply an output power of 15 kW at 1.0 PF. If the specification requires only a 0.8 PF/15-kVA UPS, an output power

TABLE 14-2: *Factory Tests for UPS Systems*

- Light load test
- Synchronization for transfer
- AC input failure and return
- Transfer to alternate source power, forward and reverse
- Four-hour heat run at rated kVA
- Voltage regulation test from 0 to 100 percent rated output with DC voltage varying from 105 V DC to 140 V DC
- Dynamic voltage regulation test to ensure transient voltage response
- Total output harmonic distortion test
- Output frequency-variation test over the full range of DC and AC bypass input voltages from 0 to 100 percent rated output
- Operational test of static transfer switch at rated output, including synchronizing circuitry
- Bypass transformer tests per ANSI/IEEE Section C57.12.01[7]
- Efficiency test at rated output if "design" test data are not available

of only 12 kW will be furnished over the range of 0.8 to 1.0 PF. As a consequence, the UPS may be initially undersized for the load it is to supply.

The specification should provide that the vendor supply an efficiency calculation. Some vendors quote efficiency of the UPS system in different terms than others. If efficiency is not evaluated, the inferior equipment may be selected.

Project owners and engineers should require enough information from bidders to ensure that any battery system supplied with the UPS is capable of the number of discharges and discharge time duration required by the specification. The vendor should submit the battery discharge curves. Hydro plant owners and engineers should be very cautious of sealed, maintenance-free batteries, owing to the failure record of such equipment. In general, several factors encourage a high failure rate: thermal runaway caused by placing sealed maintenance-free batteries in enclosures; non-temperature compensated battery chargers that overcharge the batteries; and lack of maintenance.

REFERENCES

[1] "Recommended Practice for the Application and Testing of Uninterruptible Power Supplies for Power Generating Stations," American National Standards Institute(ANSI)/Institute of Electrical and Electronics Engineers (IEEE) Standard 944, ANSI/IEEE, 1986 (reaffirmed 1996).

[2] "Emergency and Standby Power Systems for Industrial & Commercial Applications," ANSI/IEEE Standard 446 (IEEE Orange Book), ANSI/IEEE, 1987.

[3] Platts, John, and John St. Aubyn, editors, *Uninterruptible Power Supplies,* Peter Peregrinus Ltd., London, 1992.

[4] McPortland, John F. III, and John M. McPortland, "Evaluating Single Module UPS Technology: Static vs. Rotary," *EC&M,* June 1988.

[5] "Recommended Practice for Design of Reliable Industrial & Commercial Power Systems," ANSI/IEEE Standard 493 (IEEE Gold Book), ANSI/IEEE, 1980.

[6] "IEEE Standard Criteria for Class 1E Power Systems for Nuclear Power Generating Stations," ANSI/IEEE Standard 308, ANSI/IEEE, 1991.

[7] "General Requirements for Dry-Type Distribution and Power Transformers Including Those with Solid Cast and/or Resin-Encapsulated Windings," ANSI/IEEE Standard C57.12.01, ANSI/IEEE, 1979.

Chapter 15
Bus Systems

Many modern hydroelectric generators operate at substantially larger ampacities than those installed in the 1960s or earlier. The large ampacity output necessitates using buswork for the electrical connection from the generator to the step-up transformer, whereas cable was generally adequate for older systems.

The bus system equipment for a large MVA hydroelectric generator normally consists of the following components:

- A main generator bus duct connecting the generator output terminals to the low-voltage side of the step-up transformer; and
- An auxiliary bus duct connecting the generator to the unit auxiliary transformer, excitation cubicles, potential transformer cubicles, and surge protection cubicles. (The potential transformer and surge protection cubicles are utilized, respectively, for metering and protective relays and for generator surge protection.)

Standards for Bus Systems

The standards for bus systems usually are incorporated in switchgear assembly standards. In 1969, the ANSI/IEEE standard for bus was published in American National Standards Institute (ANSI)/ Institute of Electrical and Electronics Engineers (IEEE) Standard C37.20, which primarily addressed switchgear assemblies. [1] When the ANSI/IEEE standards were revised in 1987, ANSI/IEEE Standard C37.23 became the reference standard for bus equipment.[2] The pertinent International Electrotechnical Commission (IEC) standard for bus systems is IEC Standard 298.[3] Standard 298 applies to both bus and high-voltage metal-enclosed switchgear.

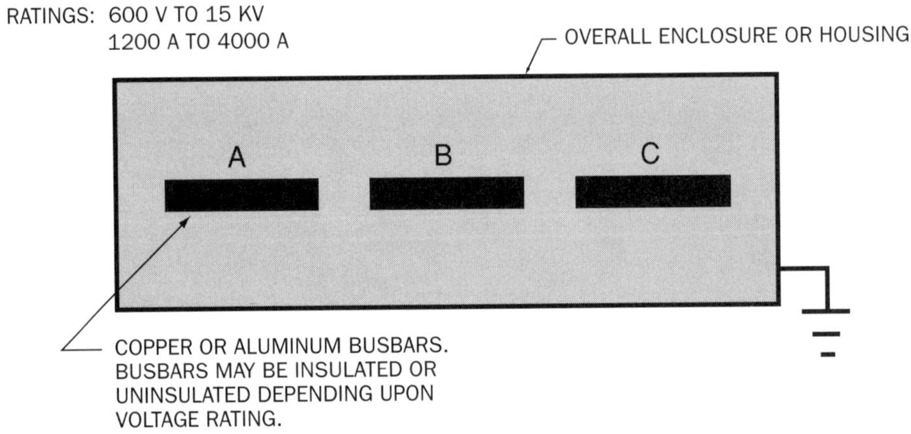

FIGURE 15-1: ***Non-segregated Phase Bus.*** *Non-segregated phase bus, normally used for relatively low voltages and ampacities, consists of three bus bars, one for each phase in a common grounded metal housing. There are no phase barriers.*

CHOOSING BUS SYSTEMS FOR A HYDRO PLANT

Types of Bus

Bus systems consist of rigidly connected conductors situated inside an enclosure, or housing. The conductors are fastened to their termination points by flexible conducting straps. The flexible straps allow the bus conductor to expand and contract with temperature changes during unit operation. The flexible straps also isolate the bus system from any vibrations occurring in the rotating generator assembly. The housing provides protection to the conductors, preventing contamination of the insulation by dust and moisture.

ANSI/IEEE Standard C37.23 defines three types of metal-enclosed bus structures: non-segregated phase bus, segregated phase bus, and isolated phase bus.[2] Non-segregated phase bus includes three bus bars, one for each phase, in a single grounded housing. There are no phase barriers in the non-segregated system. Segregated phase bus also consists of three bus bars in a single housing, but with grounded phase barriers between the bus bars. The segregated phase bus handles higher ampacities and provides a greater degree of physical protection against a phase-to-phase short circuit than the non-segregated phase bus.

In isolated phase bus, each bus bar is housed in a separate grounded metal enclosure. This type of bus provides more physical protection against phase-to-phase short circuits than either the non-segregated or segregated phase bus. Isolated phase bus normally is used when ampacities exceed 4,000 A to eliminate the possibility of induced circulating currents and to prevent a phase-to-phase fault. An additional feature of isolated phase bus is that each phase's enclosure is short

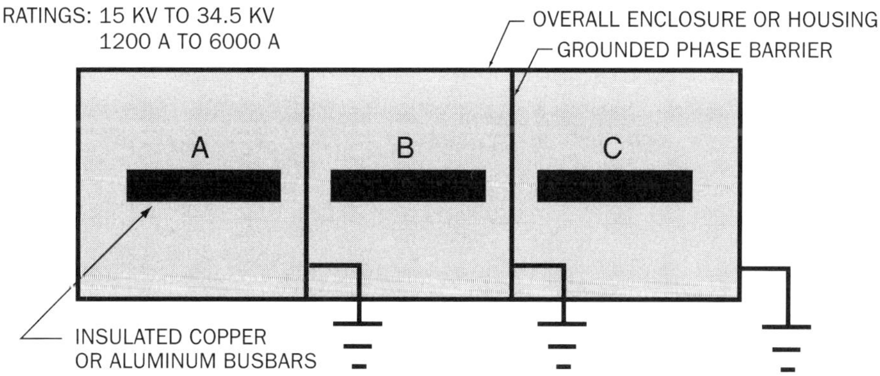

FIGURE 15-2: ***Segregated Phase Bus.*** *In segregated phase bus, three bus bars are housed in a single grounded enclosure, but are separated by phase barriers. The phase barriers provide protection against phase-to-phase short circuits.*

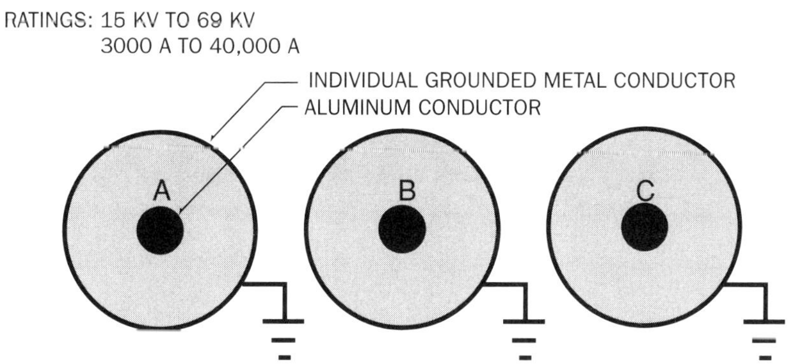

FIGURE 15-3: ***Isolated Phase Bus.*** *In isolated phase bus, each bus bar is housed in a separate grounded enclosure, providing protection against phase-to-phase faults circuits and induced circulating currents. Isolated phase bus is used for relatively high voltages and ampacities.*

circuited at the generator and transformer ends. This eliminates magnetic fields outside the enclosure and prevents any induced circulating currents in adjacent metallic structures.

Voltage, Ampacity Ratings

Voltage and ampacity ranges normally associated with the three types of bus are summarized in Table 15-1. These ranges overlap one another to some degree. Consequently, an engineering decision must be made, taking into account features of the specific hydro plant, to select the most appropriate type of bus duct. Normal practice is not to exceed 2,500 A with non-segregated phase bus, or 4,000 A with segregated phase bus. Observing these limits helps prevent the formation of extraneous magnetic fields that induce circulating currents.

TABLE 15-1: *Typical Ratings for Bus Systems*

Bus Type	Voltage Range (kV)	Ampacity Range (A)
Non-Segregated Phase Bus	0.6-15	1,200-4,000
Segregated Phase Bus	15-34.5	1,200-6,000
Isolated Phase Bus	15-69	3,000-40,000

Temperature Ratings

The various types of bus duct are rated for an ambient temperature of 40° Celsius with a bus bar temperature rise of 65° Celsius, in accordance with ANSI/IEEE Standard C37.23. [2] The maximum allowable temperature of the external bus enclosure housing is 80° Celsius if the enclosure is accessible by personnel, or 110° Celsius if it is inaccessible.

Non-segregated and segregated phase bus are physically sized using either aluminum or copper conductors to accommodate their largest ampacity ratings without any means of forced cooling. Isolated phase bus conductors, which are always made of aluminum, can be sized for normal convective cooling to an ampacity of approximately 25,000 A. For ampacity ratings from 25,000 A to 40,000 A, forced air cooling must be used to maintain the temperature rise of the bus duct and enclosure within ANSI-specified ranges.

Testing Bus Equipment

Factory and field tests for bus equipment are relatively straightforward. Design tests are performed to verify the thermal capabilities and dynamic short circuit strength of the bus duct. The factory tests consist of dielectric tests of the various bus sections in accordance with ANSI Standard C37.23. After field assembly, the bus duct joints are tested for watertightness. Then another set of dielectric tests is performed on the entire bus assembly.

Additional equipment components associated with the bus equipment, such as potential transformers and surge arrestors, are tested in the factory and the field in accordance with ANSI/IEEE standards pertinent to the component in question. All factory and field tests for the bus equipment and related components should be detailed in the equipment technical specification.

References

[1] "Switchgear Assemblies Including Metal Enclosed Bus," ANSI/IEEE Standard C37.20, American National Standards Institute (ANSI)/Institute of Electrical and Electronics Engineers (IEEE), 1969.

[2] "Guide for Metal-Enclosed Bus and Calculating Losses in Isolated Phase Bus," ANSI/IEEE Standard C37.23, ANSI/IEEE, 1987.

[3] "High Voltage Metal-Enclosed Switchgear and Controlgear," IEC Standard 298, International Electrotechnical Commission, 1969.

Chapter 15

Chapter 16
Low-Voltage Switchgear for Hydropower Plants

Low voltage (LV) metal-enclosed switchgear typically is used to distribute station service power within hydroelectric power plants that have substantial auxiliary loads. In such applications, the switchgear can provide an effective, safe, and economic means for ensuring continuity of station service. Plant owners and engineers can choose from among a number of alternatives in selecting LV switchgear for a specific hydro facility. To ensure the right equipment is chosen, the owner or engineer must carefully evaluate the alternatives available and the specific requirements of the plant's operation.

Using LV Switchgear at Hydro Plants

In general, LV switchgear is appropriate for hydro projects with powerhouses of greater than 25 MVA capacity and particularly for plants with multiple units. The switchgear falls under the American National Standards Institute (ANSI) standards for metal-enclosed switchgear.[1] This switchgear includes power circuit breakers capable of handling voltages from 240 to 600 V and consisting of removable circuit breakers with a bare primary bus. Plants with a limited number of auxiliary loads typically use power distribution panelboards with molded case circuit breakers (MCCBs) and motor control centers to feed auxiliary loads. Figure 16-1 compares a standard hydro plant electrical arrangement with and without LV switchgear.

LV metal-enclosed switchgear has several advantages over MCCBs. The switchgear is constructed for repetitive opening and closing and has higher short-circuit interrupting and momentary ratings, thus providing more flexibility and durability. LV switchgear also is constructed for a greater number of short-circuit operations between maintenance periods. Maintenance is relatively easy, owing to the easy access provided by the drawout-type circuit breakers.

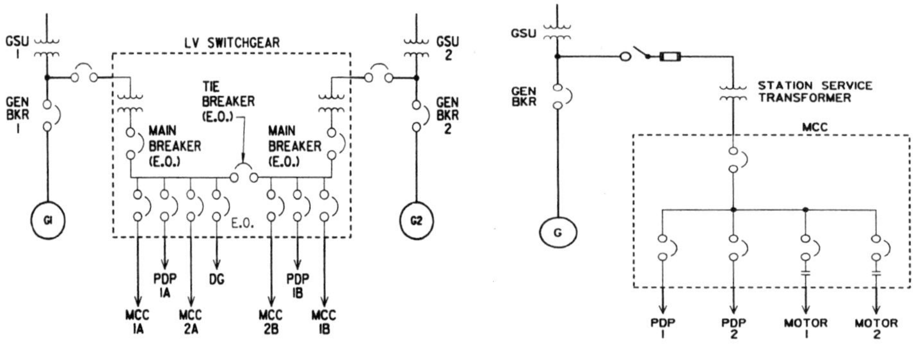

FIGURE 16-1: ***Distributing Auxiliary Power.*** *Low-voltage, metal-enclosed switchgear commonly is used in hydroelectric power plants with many auxiliary loads to distribute station service power to the turbine-generators and station auxiliary equipment. The schematic at left is a typical configuration for such a plant. Plants with limited auxiliary loads commonly use power distribution panelboards with molded case circuit breakers and motor control centers to feed auxiliary loads. The righthand schematic shows this configuration.*

Selective Coordination

The circuit breakers in LV switchgear also have integral protective devices for sensing short circuits, overcurrents, and ground faults. This provides the capability of selectively coordinated trip setpoints for each breaker. Selective coordination defines the time period in which the circuit breakers are allowed to operate. The breaker closest to the fault is the first interrupting device to operate. The next breaker upstream from the fault delays opening until the breaker nearest the fault has had sufficient time to operate. If the fault remains on the system longer than the specified time delay period, the upstream breaker immediately opens.

Selective coordination minimizes the amount of equipment out of service due to a fault. It also provides an adequate hierarchy of backup protective devices that will function in the event the device closest to the fault fails to operate. MCCBs typically have been designed for use in non-essential load circuits that do not require extremely selective coordinated tripping. MCCBs have adjustable magnetic trip elements for the short-circuit current. However, these trip elements are not as accurate as the protective devices furnished with LV switchgear circuit breakers. MCCBs can be coordinated for tripping: the smaller feeder breaker with a lower trip rating is located downstream on the circuit; the larger main breaker with a higher trip rating is located upstream. However, MCCBs were not capable of being selectively coordinated until very recently.

Using microprocessor technology, MCCBs can be provided with a function called "zone selective interlocking." This function simulates the time delay characteristics of LV switchgear by sending a signal over communication cables to restrain the upstream MCCB from tripping until the downstream MCCB closest to the fault has had the necessary time to trip. Such MCCBs are more typical in highrise buildings or industrial plants than in hydroelectric power plants.

Incorporating Switchgear in the Substation

The metal-enclosed assembly of LV switchgear usually is an integral part of a single-ended or double-ended unit substation. A single ended unit substation includes one station service transformer in conjunction with the LV switchgear. A double-ended unit substation includes two station service transformers in conjunction with the LV switchgear. A transformer is located on each side of the LV switchgear.

Selecting a unit substation where all equipment has been assembled by one manufacturer has two primary advantages over purchasing switchgear and transformer components separately. First, engineering of individual components has been coordinated to the manufacturer's standard design for size, ampacity, circuit-breaker interrupting capability, and other variables. Second, the substation has a better appearance, owing to the fact that all equipment components are by a single manufacturer and integrated into an overall assembly.

LV Switchgear Operation

Circuit Breaker Arrangement

LV switchgear consists of a series of compartmentalized, draw-out, air-interrupter circuit breakers isolated from each other and the main power bus by metal barriers. The main power bus is segregated from the circuit-breaker compartment by a metal barrier and is in the rear of the switchgear enclosure. The main power bus has no insulation cover. It consists of bare copper bars with an appropriate air spacing to prevent flashovers. The switchgear breaker is connected to this main bus by firmly pushing it back along the sidewall tracks of the breaker compartment until the main breaker stabs are engaged by the main power bus.

Each circuit breaker in the assembly can be drawn out from the connected position to test or disconnected positions. The disconnected position commonly is called the racked-out position. This denotes the fact that both the main breaker power stabs and the secondary control circuitry of the breaker are disconnected from the stationary components of the metal enclosed switchgear cubicle.

In the intermediate test position, the breaker's primary power stabs are disconnected from the stationary power bus in the rear of the cubicle, but the breaker secondary control circuitry remains connected. Consequently, the breaker may be closed and tripped as many times as necessary in order to test the breaker or to check the calibration of the integral overcurrent settings of the breaker. A mechanical interlock is provided on all of the circuit breakers. This prevents the breaker from being moved inadvertently to the closed position from the test or disconnect positions.

Tripping Options

All 480-V alternating current (AC) circuit breakers in hydro plant switchgear trip automatically to interrupt the current when a short circuit is detected. Both electrically operated and manually operated circuit breakers rely on stored-energy operating mechanisms to perform this tripping operation. However, these mechanisms are charged differently in the two breaker types.

Electrically Operated Circuit Breakers. Electrically operated, or E.O., breakers have a push button on the front that can be used to manually trip the breaker. Electrically operated circuit breakers also can be tripped remotely. A spring charging motor is automatically operated from a reliable remote power source after each breaker operation to ensure the breaker's stored-energy mechanism is fully charged. In hydropower plants, this reliable remote power source usually is 125-V direct current (DC) power from the station batteries. Other power sources (such as 48-V DC, 250-V DC from an auxiliary battery source, or 120- or 240-V from AC essential service) may be used, depending on their availability in the powerhouse.

A capacitor-stored energy trip device may also be purchased with the breaker. The capacitor provides energy to trip and close the breaker, if normal essential power is not available in the plant when a tripping operation is required.

Manually Operated Circuit Breakers. Manually operated, or M.O., circuit breakers are opened and closed manually by an operator standing in front of the breaker cubicle, similar to manual operation of an E.O breaker. However, the operator then must recharge the stored-energy mechanism of the M.O. breaker by pumping the manual recharging handle located on the front of the breaker cubicle. Once the stored-energy mechanism is recharged, the breaker may be closed in anticipation of the next tripping operation. M.O. breakers may be furnished with an optional shunt trip device to allow remote tripping of the breaker. However, an operator must return after a remote trip of an M.O. breaker to the front of the breaker to manually recharge the breaker's stored-energy mechanism before the breaker can be closed again.

M.O. breakers cost approximately 30 percent less than comparably sized E.O. breakers. This initial cost savings usually means that M.O. breakers are used for all switchgear feeder breakers and other non-essential functions in hydro plants. For a plant with a double-ended substation, E.O. breakers would be used only for the two main incoming feeds, the bus-tie breaker, and possibly the emergency diesel generator connecting breaker.

Current-Limiting Fuses

Manufacturers offer LV switchgear with current-limiting fuses as an integral part of the switchgear. The current-limiting fuses are series connected to the circuit breaker, and increase the short-circuit interrupting capability of the switchgear while maintaining the same breaker frame size. Switchgear with current-limiting fuses is used when the available fault currents at the LV switchgear location exceed the symmetrical interrupting current of the standard circuit breaker frame size.

Current-limiting fuses mounted integrally in the switchgear allow for a more physically compact switchgear at a lower cost, while providing the short circuit interrupting capability required. However, the fuses must be replaced after every short-circuit event, requiring the owner to keep a quantity of spare fuses on hand. This type of LV switchgear is very common in industrial plants where numerous plant expansions have occurred or additional equipment has been added to the plant since initial installation. Fused switchgear is not very common in hydroelectric plants.

Breaker Ratings

The circuit breaker frame size determines the maximum momentary current and symmetrical interrupting currents of the breaker. The momentary, or asymmetrical, current is the high initial in-rush of current to ground when a short circuit first occurs. Commonly called the root mean square (RMS) asymmetrical fault current, this current reflects the closing and latching capability of the circuit breaker. Once the short circuit has become well established, the RMS symmetrical fault current is the current that the circuit breaker must be capable of interrupting to stop the flow of current to the area of the short circuit.

Breakers of a given frame size may be purchased with a wide variety of trip ratings. For example, a breaker with an 800-A frame may be found with trip ratings in a range of increments from 100 to 800 A. The plant owner or engineer should select the trip rating based on the actual load current that will be supplied from the breaker. The trip rating of a breaker is directly related to the available long-time, short-time, instantaneous, and ground fault protective setting ranges. These ranges are calibrated in fractions or multiples of the trip rating value selected.

Example: Selecting a Circuit Breaker

As an example, consider an application in which a circuit breaker must be selected to supply 630 A to various loads in a hydropower plant. The system RMS symmetrical fault current has been calculated as 35,000 A at 480 V. The RMS asymmetrical, or momentary, fault current has been calculated as 56,000 A. The first step in selecting an appropriate breaker is to find one with a frame size capable of interrupting the symmetrical short-circuit current and withstanding the asymmetrical fault current. Using a typical chart of breaker capability such as shown in Table 16-1, a plant engineer or owner would choose the model with the 1,600-A frame size.

The next step is to select the trip rating for the unit that is adequate for the required load of 630 A. In this typical example, the 1,600-A frame size has available trip ratings of 200, 300, 400, 600, 800, 1,200, or 1,600 A. The rating of 800 A should be selected.

Based on that trip rating, a number of settings are available for long-time, short-time, instantaneous, and ground-fault sensing. Table 16-2 shows a typical range. A coordination study based on the system short-circuit parameters must be performed to identify the exact setting to be selected. The selected setting must be coordinated with any upstream and downstream circuit breakers and fuses to provide selective tripping. The circuit breaker immediately upstream should be coordinated, or selectively set, to trip next in the event that the breaker nearest the fault fails to operate. This sequence should be continued so that all circuit breakers in the powerhouse are coordinated.

Protective Functions

LV switchgear circuit breakers are furnished with solid-state microprocessor-based electronic trip units with long-time, short-time, instantaneous, and ground fault setpoint ranges. Each of the ranges must be selectively coordinated for the load it feeds. Though the appropriate settings vary slightly with different manufacturers, several general rules apply.

Table 16-1: *Low-Voltage Switchgear Frame Sizes and Characteristics[2]*

Breaker Type Designation	Available Frame Sizes	480-V RMS Symmetrical Interrupting Current (A, RMS)	480-V RMS Asymmetrical Fault Current (A, RMS)*
DS-206	800-A	30,000-A	48,000
DS-416	1,600-A	50,000-A	80,000
DS-420	2,000-A	65,000-A	104,000

*Per ANSI/IEEE Standard 141, the asymmetrical fault current is 1.6 times the symmetrical interrupting current.

Table 16-2: *Setting Ranges for Circuit Breaker Protection Mechanisms[2]*

Protection	Setting Range in Multiples of Breaker Normal Current	Time Range of Operation
Long Time	0.25 to 1.25	4 to 36 seconds
Short Time	2.0 to 10.0	11 to 30 Hz
Instantaneous	2.0 to 12.0	3 to 5 Hz (approx.)
Ground Fault*	180A	1 second (approx.)

*Ground fault sensors usually have selectable ranges. The minimum range setting usually is selected in order to detect a ground fault as early as possible.

Long-Time Settings

Long-time settings are used to provide supplemental overload protection for equipment such as transformers and motors. Such overloads occur over a relatively long time (four to 30 seconds). Long-time breaker settings typically are 125 percent of a motor's full load amperes to prevent motor burnout.[3] For transformers, the secondary side circuit breaker can be set at 125 percent of the maximum current output to avoid transformer damage. If this 125 percent overcurrent value is maintained for the long-time setting time period of four to 30 seconds, the circuit will detect any abnormal condition and trip.

Short-Time and Instantaneous Settings

Short-time settings usually range over a time period of 11 to 30 cycles while the instantaneous time of operation is from 3 to 5 cycles. The short-time setting indicates that significant overcurrent is present and that the breaker should trip more rapidly than it would for a long-time trip condition. Short-time setting ranges are in multiples of two to ten times the breaker's normal rated current. The instantaneous setting range is two to 12 times the normal rated current.

The short-time setting is designed to delay tripping of the breaker even though a major problem has occurred. This delay will allow circuit breakers closer to the fault adequate time to trip. The instantaneous setting detects a fault closer to the breaker and requires the breaker to trip immediately.

Ground Fault Settings

Integral ground fault pick-up settings usually range from 20 to 60 percent of the ground sensor rating. For most equipment, the ground sensor rating equals the continuous current rating of the breaker. As a consequence, even with a 20 percent minimum range selected, a standard 2,000-A frame circuit breaker with a continuous current rating of 2,000 A would have a minimum ground fault setting of 400 A. This is a substantial amount of current going to ground before the integral ground protection of the breaker is able to sense the problem. Therefore, many hydro plant owners supplement the integral breaker ground fault circuit with a separate, more sensitive protective relay circuit to detect ground faults.

Other Capabilities

New microprocessor-based smart trip units have numerous capabilities in addition to long-time, short-time, instantaneous, and ground fault protective functions. Among other capabilities, the new units can communicate digitally with programmable logic controllers, computers, and other digital systems in the plant. This capability allows the plant computers to process alarm outputs, receive self-monitoring trouble outputs, and monitor event-tagging information received from the circuit breakers. A number of optional features may also be purchased with each switchgear assembly. Among the more attractive are:

- Operation counters;
- Shunt trip attachments that allow remote tripping of M.O. breakers;
- Capacitors to charge the stored-energy operating mechanism of an E.O. circuit breaker when it is remotely located; and
- Compartment position, or truck-operated cell, switches that indicate the position of the breaker and can be used to bypass a switchgear auto transfer scheme when one of the E.O. main incoming breakers is in the "test" position.

Upgrading LV Switchgear

Retrofitting of LV switchgear circuit breakers installed from the 1950s to the 1970s with solid-state trip systems has become quite common. The retrofit of the breakers' protective trip system must be coordinated with a complete breaker overhaul and test program to ensure that the reliability of the older technology breaker is maintained. The retrofit/breaker overhaul should be performed by a firm that has performed this type of work for at least 10 years, has solid client references, and will warranty the work performed for at least 2 years.

Retrofitting of older technology LV switchgear circuit breakers will provide the following benefits:

- Improved coordination with upstream and downstream devices;
- Better reliability and elimination of nuisance trips; and
- Digital metering and communication capabilities.

The retrofit or overhaul of LV switchgear breakers should be performed in conjunction with a system coordination study. The system coordination study determines the short circuit ampacities at various points in the electrical system and coordinates the breaker trip settings with upstream and downstream devices to ensure a selectively coordinated system is achieved.

Due to the volume market that has occurred for retrofitting and reconditioning older technology circuit breakers, ANSI has developed a conversion standard (ANSI/IEEE Standard C37.59) that addresses service conditions, ratings, the conversion process, and design verification testing.[4] This standard should be applied rigorously to ensure a retrofit/overhaul installation of LV switchgear circuit breakers proceeds smoothly.

Selecting a Supplier

Historically, primary circuit breaker manufacturers such as General Electric, Westinghouse, Siemens, and ABB were the only sources of LV switchgear assemblies for hydropower plants. However, since the early 1980s the LV switchgear assemblies have also been offered by companies that purchase circuit breakers from the

FIGURE 16-2: *Substation Options.* *To avoid engineering problems and achieve an attractive, integrated appearance, owners should select unit substations in which all equipment has been assembled by a single manufacturer. The double-ended substation shown above, while functional, uses switchgear and transformers from different manufacturers. The switchgear is in the center of the photo, flanked by two oil-filled transformers. Due to manufacturer differences, this configuration requires an interconnecting bus duct above the switchgear and transformers. This contrasts with a double ended substation in which all the equipment is provided by a single manufacturer, such as the one shown in the photograph at right. In such substations, the components typically have been engineered to be fully compatible.*

major manufacturers and fabricate the LV switchgear assembly. Because of their lower overhead, fabricators sometimes can offer a price savings of 20 to 40 percent for a complete unit.

The issue is identifying a reputable fabricator. Some fabricators have limited quality assurance and compliance programs. Many fabricators have limited drawing production facilities and lack experienced staff. This can result in drawing submittal delays, inadequate checking and review of the drawings at the factory, and a poor quality switchgear assembly. These problems are difficult to distinguish in the bidding phase of a switchgear contract, and their extent typically becomes evident only during the factory inspections or after the equipment is shipped to the field. Whenever discovered, such problems can more than offset the initial savings realized by purchasing from a switchgear fabricator.

The solution lies in two approaches: accepting bids only from reputable fabricators whose equipment has been previously used and proved satisfactory; or using the bid specifications to restrict the bidders to major firms that also manufacture the circuit breakers.

References

[1] "IEEE Standard for Metal-Enclosed Low Voltage Power Circuit Breaker Switchgear," Institute of Electrical and Electronics Engineers (IEEE) Standard C37.20.1, IEEE, 1993.

[2] *Westinghouse Consulting Application Guide*, Westinghouse Catalog 55-000, 1991-1992.

[3] "Recommended Practice for Protection and Coordination of Industrial and Commercial Power Systems," American National Standards Institute (ANSI)/IEEE Standard 242 (Beige Book), 1991.

[4] "IEEE Standard Requirements For Conversion of Power Switchgear Equipment," ANSI/IEEE C37.59, ANSI/IEEE, 1991.

Bibliography

Brown, J. Guthrie, *Hydro-Electric Engineering Practice,* 2nd Edition, Blackie and Son Ltd., Glasgow, Scotland, 1970.

Smeaton, R.W., *Switchgear and Control Handbook,* Second Edition, McGraw-Hill Book Co., New York, 1987.

Chapter 17
Start-up and Commissioning of a Hydroelectric Project

Comprehensive start-up and commissioning procedures for hydroelectric plants are not well defined in standards presently issued by the National Electrical Manufacturers Association (NEMA) , the American National Standards Institute (ANSI) and Institute of Electrical and Electronics Engineers (IEEE), or the International Electrotechnical Commission (IEC). The standards commonly used in the field during commissioning are related to commissioning or installation of the turbine or generator equipment components.[1,2,3,4,5] The turbine/generator commissioning standards, however, do not address any start-up tests or procedures for the remaining station and unit auxiliary electrical systems that are an essential part of a hydropower plant.

Various terms are used to describe the different phases of the process, and often have different meanings to equipment vendors and owners. Such loosely-defined terms include start-up, pre-start-up, performance, field check-out, unwatered, watered-up, pre-commissioning, and commissioning tests. The lack of precision in naming the various phases of start-up and commissioning also results in different equipment tests being performed in different phases, depending upon how the contractual specification was written and interpreted.

IEEE Standard 1248, issued in 1998, was intended to define the entire sequence of commissioning for the "electrical" systems in a hydroelectric plant, including the station and unit auxiliary electrical systems.[6] This standard defines a typical start-up organization team, the various electrical system and equipment tests, and the approximate time durations to complete commissioning for various electrical systems. This standard is a significant step forward with respect to the start-up tests and procedures for electrical systems at a hydroelectric station. Unfortunately, no complementary mechanical standard exists.

Stages in Start-Up and Commissioning

The process of start-up and commissioning for hydroelectric plants has several stages. Although these stages may be called by different names and differ in specifics, they follow the same general sequence.

Field Check-Out (Systems Operational Testing)

This is the stage in which all systems and subsystems are individually tested. When a complete, uninterrupted system checkout is not possible, segmented system checkout is acceptable. Segmented system checkout must, however, incorporate overlapping segments to ensure that each system has been fully checked out. Tests at this stage include a check of all critical unit clearances and dimensions; alignment of the unit shaft system; calibration and adjustment of all temperature switches, pressure switches, flow switches, transmitters, etc.; continuity and megger tests of all cabling; and operation of all control systems in both the de-energized and energized states. Check-out also includes all field testing that can be performed before the turbine begins to operate, such as hydrostatic tests of all pressurized fluid systems, AC/DC high-potential tests, and bench tests of protective relays.

All check-out tests must be adequately documented for each system and subsystem. The test records must be signed by both the individual tester and the electrical or mechanical supervisor. This documentation allows the owner or engineer to identify potential problems in specific equipment before start-up, and makes it easier to determine the source of any equipment failure or malfunction.

Start-Up Tests

Start-up tests are tests of the turbine/generator unit. At the time of start-up testing, all other systems and subsystems should have been already tested during the field check-out. Start-up tests are usually divided into unwatered and watered-up tests.

Unwatered Tests (Pre-Start-Up Tests, Pre-Commissioning Tests). These tests are conducted before the unit has been watered up, and are intended to confirm that the unit will be in acceptable condition for the "watered-up" tests that follow. Most of the "unwatered" generator tests listed could be performed in parallel with the field check-out testing.

Some typical tests are hydrostatic tests of generator coolers and piping; insulation resistance and dielectric tests of the generator; functional checks and adjustments of generator speed switches, limit switches, and pressure switches; and functional checks of the brakes, wicket gates, and excitation system. In addition, immediately before first rotation of the unit, the entire unit must be thoroughly inspected for tools or other objects which may have been inadvertently left in the unit. When this inspection is complete, the unit is ready for "watered-up" tests.

Watered-Up Tests. At this stage, the unit is watered up, mechanically rotated for the first time, and synchronized to the system. The water passages are filled and the unit produces its first power. This is a critical stage because, for the first time, all of the equipment components are operating as an integral system under rated head and flow conditions. The first unit rotation is commonly called the "bump test." The unit is turned only slightly ("bumped") and the test personnel listen for unusual noises or interference contacts. After the bump test, the unit is run at 25, 50, 75, and finally 100 percent of rated speed.

The electrical tests at the "watered-up" stage include all on-line field tests of the generator, such as the telephone influence factor and waveform deviation tests, shaft runouts, and bearing temperature checks. During this stage of start-up testing, the owner/engineer confirms that he can reliably start, stop, and synchronize the unit to the system. All protective devices, lock-out relays, and emergency stop systems are checked at reduced power loads to ensure that they are functional. Unit load rejection tests also are performed, ensuring that the maximum transient inlet pressure rise does not exceed the design pressure.

The electrical tests included in the "watered-up" stage of start-up testing are usually broken down into two subcategories. The first of these categories is referred to either as no-load tests or generator initial-operation tests. No-load tests confirm the operation of the generator and powerhouse auxiliary equipment under no-load or minimal load conditions. Tests in the second category, called load tests or performance tests, confirm the operation of the generator under load conditions. After the no-load tests have been completed, the unit load is increased, usually in increments of 25 percent, until 100 percent rated load is attained. At each step, a load-rejection test is performed. The load-rejection

TABLE 17-1: ***No-Load and Load Tests Performed During Watered-Up Testing***

No-load tests (generator initial operation tests):

- Phase rotation check
- No-load saturation test
- Short circuit saturation test
- Operational tests, checks of thermal relays, speed switches, RTDs, flow switches
- Excitation system checks

Load tests (performance tests):

- Heat run tests to determine maximum temperature rise
- Load rejection tests
- Brake operational time to safely stop the unit
- Measurement of excitation field current at generator rated output
- Measurement of the cooling water flow rates for the bearings
- Unit capacity tests

tests at 25, 50, 75, and 100 percent rated load confirm that the unit can be safely stopped under any operating condition. Typical no-load and load tests are listed in Table 17-1.

During the load rejection tests, the penstock pressure transient rise and the unit overspeed are measured to confirm respectively that the pressures are not exceeding design pressures and that unit overspeeds are within design parameters. Any unusual noises, equipment misoperations, high temperatures, alarms, or other unusual or unexpected conditions that occur during the incremental load operations should be thoroughly investigated.

At the completion of this stage, the owner or engineer should be confident that the turbine/generator and all associated controls and subsystems are functioning as designed. Problems that are not visible at a reduced or short-term unit power rating, but may occur during sustained rated power output of the unit, are resolved in the final commissioning stage.

Commissioning Tests

In this stage of testing, the unit is operated at 100 percent of the manufacturer's rated output for a specified number of days. If no major problems occur after operation at full rated output for the contractually required period, the unit is accepted by the owner for commercial operation.

Unit index tests and efficiency tests are usually conducted during the commissioning phase to verify that the unit complies with all performance aspects of the contract. Some specifications are written to include certain performance tests or load tests in the commissioning phase. Others are written to include certain commissioning tests in the performance test phase. The structure of the testing in the specification depends on the historical experience of the owner and the engineer.

Testing the Control System

The individual equipment tests that should be performed on specific equipment components, such as the generator, are adequately defined by ANSI/IEEE standards.[1] However, this clarity of definition does not extend to hydroelectric plant control systems. This is because control system verification and testing is one lengthy, integrated process, whereas most equipment testing is a series of pass/fail tests.

Verification of a control system includes tests of individual components, connections, insulation, and software. Necessary tests and checks include:

- Point-to-point continuity tests to verify that the components are connected properly;
- Insulation resistance tests to check the condition of the cable insulation;
- Functional tests of each component and the interconnected system of components in both the automatic and manual modes; and
- Tests of the software and hardware associated with the computer control system.

Commercial Operation

Commercial operation begins on the date the plant is officially available to sell power to the public. This date immediately follows completion of the commissioning tests. At this time, the plant enters the rate base by agreement (if applicable) with the appropriate regulatory agency.

Ensuring Successful Commissioning

Start-up of a hydroelectric plant involves a great many components, procedures, and suppliers. The owner's vigilance and attention to clear specifications and documentation are essential in achieving a smooth start-up and timely resolution of problems that may arise.

Field Check-Out and Testing Procedures

The specifications must stipulate that all major equipment suppliers provide written procedures for the field check-out and testing of the equipment furnished. Field check-out and test sheets must be filled out in the field during the check-out and test procedure. The completed sheets should be signed by the manufacturer's test representative and the owner's electrical/mechanical supervisor. The field check-out and test sheets must be diligently filed for future reference before beginning the "unwatered" and "watered-up" start-up and commissioning tests.

Manufacturers often claim that written procedures are not required since the manufacturer's representatives supervise and sometimes perform all installation and testing. However, written procedures allow the owner, the engineer, and the general contractor to understand, coordinate, and track the work. Most importantly, they prevent field personnel from taking short cuts, deleting checks and procedures, or performing work that does not meet industry standards.

The Role of Manufacturer's Representatives

A manufacturer's representative must be available at the site for each major equipment component during check-out and field testing of the equipment. If at all possible, the same testing personnel that performed the factory tests should perform the field tests of the equipment. This provides continuity and ensures that the test personnel will be aware of any peculiarities or problems that occurred during the factory tests of the specific equipment item.

The Resident Engineer

Most hydro plant owners assign an engineer to reside at the site during the construction and installation of the powerhouse. This engineer is called the *resident engineer*. The resident engineer is responsible for interpreting the contract, responding to questions from the general construction contractors and equipment vendors, and coordinating the contract performance. The number of resident engineers at a project site depends upon the size and complexity of the project. Usually, only one resident engineer is on site for construction of a small hydroelectric project. The technical background required of the resident engineer and any assistants depends upon the project phase.

As a rule, most engineering firms and owners employ resident engineers with civil or structural engineering backgrounds. This is adequate for the initial phases of the project when the civil works (such as the powerhouse, intake gate area, and penstock) are being constructed. However, this is not an adequate background for the mechanical and electrical phases of the project. Specialized knowledge is essential to recognize errors in electrical and mechanical field testing procedures. A resident engineer lacking this knowledge must rely exclusively on the field test certificates and documented check-out tests performed by the contractors' test engineers. Although these documents are important, they cannot substitute for personal expertise in the electrical/mechanical engineering field. The resident engineer, in addition to having the proper expertise, must also be given the time to witness all field tests and check-out procedures.

Scheduling Start-Up and Commissioning

For small hydro plants, the specifications should stipulate that the lead equipment supplier (usually the turbine supplier) is responsible for start-up and commissioning activities. This means that the lead equipment supplier is responsible for furnishing a comprehensive and coordinated schedule for start-up and commissioning that chronologically details field activities. For medium and large hydro plants, where the tasks to be accomplished during start-up are many and complex, the owner should appoint a start-up coordinator. The coordinator's responsibility is to review and integrate start-up and commissioning schedules from the various equipment suppliers into a comprehensive chronological schedule.

After completion of the "field check-out" of the various electrical/mechanical systems, typical time durations for the remainder of the start-up and commissioning tests (unwatered watered-up, and commissioning) should not exceed 10 to 15 days for a small hydro installation. Typical time durations for medium and large hydro installations vary considerably, depending upon the number of units to be commissioned and the complexity of the associated engineering works (spillways, intakes, etc.) A normal time duration for larger units (excluding field check-out testing) is approximately 30 days per unit.

The comprehensive schedule should be used to coordinate start-up and commissioning activities in relation to equipment test dates and mobilization of personnel. A well-planned schedule will allow a number of field check-out tests and "unwatered" start-up tests to be performed in parallel, minimizing the time duration of start-up and commissioning. An inadequate or poorly planned start-up and commissioning schedule will result in costly delays, and possibly personnel injuries and equipment damage.

Engineering Review during the Commissioning of a Small Hydropower Plant

Technical flaws at small hydropower plants are more common than usually recognized. These technical flaws often originate with the mistaken impression that the small power output of the plant corresponds to a small amount of required engineering input. When the developer of a small project holds this misperception, the results often include a general lack of engineering detail in the project design and planning, and a specification where the major equipment suppliers are responsible for all project engineering without any external review. These conditions tend to lead to technical flaws that may not be discovered until start-up and commissioning of the unit. By this time, the flaw is usually not correctable without a large, unexpected expenditure on the owner's part.

Finding a Qualified Engineer

Each hydroelectric plant is unique in certain aspects. This applies equally to large and small plants. If proper engineering diligence is not used during the initial design stages, the owner will obtain a poorly designed plant that cannot produce the expected energy due to a lack of water, inappropriate turbine settings or net head ratings, or inappropriately sized generators. And, if proper engineering diligence is not also applied during fabrication and installation of the equipment, start-up and commissioning will become an anxiety-filled series of turbine/generator equipment problems, incorrect plant wiring, control system failures, and protective relay trips.

Many small hydro sites today are constructed as "water-to-wire" packages with equipment installation and engineering provided by the equipment vendor. In many cases, the owner has no immediate knowledge of the design and engineering associated with a hydropower facility, and is enticed by the equipment vendor to believe that the involvement of outside engineering firms is an unnecessary extra cost. However, the use of an engineering firm for external review can be invaluable. Knowledgeable engineers should be employed to check the initial design of a small hydro plant prior to authorizing the equipment vendors to proceed with fabrication. Using an external engineering firm during fabrication and installation of the equipment will provide added assurances that the plant will operate correctly and generate the output power required by the specifications.

In addition, most banks that provide the financing for a hydro project employ a hydroelectric engineering firm to separately review the final equipment specifications, the site construction progress, and the quality of construction at site. This firm's function is to assure the bank that the unit has been constructed properly and will be commissioned on time; notify the bank promptly if there are construction problems or commissioning delays; and confirm that the unit will produce its guaranteed power output. In addition, if any major problems do occur during commissioning which result in the destruction of equipment and delay of project commissioning, the bank's engineer will be available as an independent witness capable of offering his rendition of events to the bank as well as the bank's client.

The credentials of the owner's outside engineering firm and the bank's engineering firm should reflect an extensive past history of hydropower experience. This extensive hydropower experience will eliminate the possibility of substantial errors concerning design values, such as water passageway head losses, unit setting, unit speed, and guaranteed unit output, that would be catastrophic to the project schedule or project completion.

Powerhouse Technical Review

Powerhouse technical review by the owner's or the bank's engineering firm consists of reviewing or inspecting a specific list of technical items. The technical items selected for review should be based on the engineer's hydroelectric experience, and each technical item identified should be checked off as it is reviewed or inspected.

Some items require review before the startup process begins, and other parts of the review take place during start-up and commissioning actviities. Tables 17-2 and 17-3 separate the necessary review steps into pre-start-up and start-up categories.

Technical Review of Electrical Systems. When performing a technical review of electrical systems, the engineer should:

- Ensure that all factory tests and field tests required by the specifications have been performed;
- Inspect the test certificates to confirm that the recorded values are sensible, and that the test certificate has been signed off;
- Pay particular attention to megger and high-potential tests on the major items of electrical equipment (generator, main power transformer, medium voltage circuit breakers and cables);
- Review the system ground resistance test certificate, and ensure that adequate ground connections have been provided to all major items of equipment and that all cable trays, stairways, handrails, etc. have been grounded in accordance with the National Electric Code (NEC);
- Review the powerhouse/switchyard for surge protection, noting especially the surge protection provided for externally routed control cables (a good example is the headwater level/tailwater level transmitter cables);
- Confirm that a phase rotation check has been performed at the generator circuit breaker, and that there is a signed test certificate confirming the correct generator phase rotation (and, if possible, witness this test);

TABLE 17-2: ***Items to Review Before Unit Start-Up and Commissioning***

Combined Electrical/Mechanical Review Tasks:

- Review factory and field test certificates for all major pieces of equipment:
  - Turbine
  - Governor
  - Speed increaser
  - Intake, draft tube gates
  - Generator
  - Main power transformer
  - Generator switchgear (medium voltage)
  - Low voltage switchgear
  - Unit or plant control switchboard
- Confirm that manufacturers have provided detailed written installation instructions for all major pieces of equipment, including detailed dimensional record sheets.
- Review all detailed dimensional record sheets in conjunction with a visual inspection of all major pieces of equipment.

Mechanical Review Tasks:

- Verify that hydrostatic tests have been performed on all pressurized fluid systems.
- Review operational check-out sheets for each mechanical system, including calibration sheets for all level switches, flow switches, pressure switches, etc.

Electrical Review Tasks:

- Review system ground resistance test certificate and ensure adequate ground connections per NEC.
- Review surge protection for powerhouse/switchyard.
- Confirm that a phasing check has been performed across the generator circuit breaker.
- Review the station battery/battery charger arrangement and operational check-out sheet.
- Verify that all protective relays were "bench-tested," and that the settings are acceptable to the connecting utility.
- Review PLC coding documentation and confirm that the most recent copy is available at site.

TABLE 17-3: ***Items to Review or Inspect During Unit Start-Up and Commissioning***

Mechanical Review Tasks:

- Observe first mechanical rotation of unit.
- Observe and/or review results of bearing temperature runs.
- Observe and/or review results of unit load rejection tests
- Observe turbine index tests, verify that results are consistent with manufacturer's promised performance.

Electrical Review Tasks:

- Observe unit manual starts, stops, and synchronizations.
- Observe unit automatic starts, stops, and synchronizations.
- Observe one manually initiated emergency stop sequence.
- Observe unit stops by activation of each lockout relay.

- Review the station battery and battery charger(s) arrangement - two battery chargers are preferable to one, although one is acceptable;
- Confirm that each battery charger has a test certificate verifying that the "float" and "equalize" potentiometers were set according to the manufacturer's recommendation;
- Verify that the protective relays were "bench-tested," that the relay settings are acceptable to the connecting utility, and that the CT shorting screws have been removed;
- Witness a number of manual and automatic start/stop sequences;.
- Witness a generator circuit breaker trip by activation of each lockout relay; and
- Witness one manually initiated emergency stop sequence.

For plants where a programmable logic controller (PLC) is used for unit control, some additional review items apply:

- Identify the manufacturer's recommended chassis and logic ground points, and determine whether they are required to be connected together or grounded separately. If they will be connected together, confirm how common mode noise is prevented from entering the PLC;
- Verify that the ground resistance values for the chassis and logic grounds are consistent with the manufacturer's recommendations;
- Ensure that surge protection is provided on all inputs to the I/O racks that would be susceptible to voltage transients or spikes, including, at a minimum, all cables routed to devices located outside of the powerhouse;
- Ensure that the 120-VAC power for the PLC is reliable and free of voltage transients. This is usually accomplished by feeding an inverter from the 125-VDC station batteries, which in turn provides 120 VAC to the PLC; and
- Confirm that the final PLC software coding has been well documented, and that the documentation is available for future plant maintenance and trouble shooting.

Technical Review of Mechanical Systems. To perform a technical review of the plant's mechanical systems during construction, the engineer should:

- Verify that all pressurized fluid systems (for example, hydraulic oil and cooling water) have been hydrostatically tested, and review the operational check-out sheets for each of these mechanical systems;
- Witness the unit's first mechanical rotation and the bearing temperature break-in runs (the first operational tests of the bearings under design conditions);
- Verify unit alignment and balance;
- Monitor for excessive vibration and guide bearing runout;

- Witness the unit load rejection tests to confirm that the maximum transient inlet pressure is below the penstock design pressure, and to determine the maximum transient overspeed of the unit; and
- Witness the turbine index tests to verify that the unit can attain the manufacturer's guaranteed efficiencies.

Finally, the engineer should ensure that the generator/speed increaser/turbine manufacturer has provided written installation instructions which include detailed dimensional record sheets. These sheets are used to record important installation information—for example, the installed generator air gap dimension, the installed turbine runner clearance, and the installed turbine-generator-speed increaser shaft alignment. The completed dimensional record sheets should be compared with the generator/turbine manufacturer's tolerance requirements to confirm proper installation of the equipment. They will also be used during future unit maintenance outages to identify problems such as misalignments.

References

[1] "Test Procedures For Synchronous Machines," Institute of Electrical and Electronics Engineers (IEEE) Standard 115, IEEE, 1995.

[2] "Guide for Commissioning, Operation, and Maintenance of Hydraulic Turbines," International Electrotechnical Commission (IEC) Standard 545, IEC, 1976.

[3] "Field Acceptance Tests to Determine the Hydraulic Performance of Hydraulic Turbines, Storage Pumps, and Pump-Turbines," IEC Standard 41, IEC, 1991.

[4] "Installation of Vertical Hydraulic Turbine-Driven Generators and Reversible G/M For Pumped Storage Installation," National Electrical Manufacturers Association (NEMA) Guide 5.2, NEMA, 1989.

[5] "Guide for Commissioning, Operation, and Maintenance of Storage Pumps and Pump-Turbines Operating as Pumps," IEC Standard 805, IEC, 1985.

[6] "IEEE Guide for the Commissioning of Electrical Systems in Hydroelectric Power Plants," IEEE Standard 1248, IEEE, 1998.

Sources

Several articles authored by Mr. Clemen and originally published in *Hydro Review* or *HRW* were updated and adapted for this book. These articles and the book chapters in which they appear are listed below.

Chapter 3
"Hydro Generators: Determining the Right Electrical Factory and Field Tests," *Hydro Review*, Volume 8, No. 6, December 1989.

Chapter 4
"Safeguarding Your Generators with Scheduled Inspection Outages," *Hydro Review*, Volume 17, No. 1, February 1998.

Chapter 6
"Testing: The Key to Trouble-Free Transformers," *Hydro Review*, Volume 10,
No. 4, July 1991.

"Obtaining the Right Large Power Transformer for a Hydro Plant," *Hydro Review*, Volume 14, No. 4, July 1995.

Chapter 7
"Implementing an Effective Transformer Maintenance Program," *Hydro Review*, Volume 16, No. 4, August 1997.

Chapter 8
"Q&A: Testing Large Power Transformers," *HRW*, Volume 2, No. 2, Summer 1994.

Chapter 9
"Enhancing Fire Safety at Hydro Plants with Dry Transformers," *Hydro Review*, Volume 12, No. 4, June 1993.

Chapter 11
"Testing Unit Control Switchboards In the Factory and Field," *Hydro Review*, Volume 9, No. 6, December 1990.

Chapter 12
"Are You Prepared? Protecting Your Hydro Plant from Lightning Strikes," *Hydro Review*, Volume 12, No. 7, December 1993.

"Protecting Modern Hydro Plant Controls from Electrical Surges," *Hydro Review*, Volume 11, No. 2, April 1992.

Chapter 13
"Charging Up Your Knowledge: Emergency Battery Systems for Hydro Plants," *Hydro Review*, Volume 11, No. 6, October 1992.

Chapter 14
"Choosing an Uninterruptible Power Supply for a Hydro Plant," *Hydro Review*, Volume 13, No. 4, June 1994.

Chapter 16
"Choosing Low-Voltage Switchgear for Hydro Plants," *Hydro Review*, Volume 15, No. 4, June 1996.

Index

acceptance tests, batteries 13-11
air terminal
  Franklin type 12-3,4,5,7,8,9
  spherical type 12-7
alternative (non-oil) liquid-immersed transformers 9-1
ampacity 2-10
applied voltage test 6-3,13; 9-6
architecture, control systems
  open 10-8,9
  proprietary 10-8,9
armature voltage 2-10
Askarel 9-4
assembly, generator 3-1,3,4,5,6,10,11,12
automatic synchronizing 11-3,6
auxiliary power distribution 16-1,2
auxiliary power systems 9-1,2
auxiliary transformers 5-3,4,5
basic impulse insulation level 9-3
batteries 13-1,2,4,5,6,7,8,9,10,11; 14-3,8,10
battery chargers 13-1,2,7,8,9,10
  automatic switching 13-9
  quantity required 13-3
  temperature-compensated 13-8,9
  voltage regulation 13-8,9
battery life 13-9
battery maintenance 13-10
battery monitoring system 13-10
battery rooms 13-4,5,6
battery testing 13-11
bearings
  jack screw type 2-7
  Kingsbury 2-7
  leveling plate self-equalizing 2-7
  Michell 2-7
  self-compensated diaphragm type 2-7
  spring bed mattress type 2-7
  swivel type 2-7
  tilting pad 2-7
bonding 12-3,5,10,15
braking systems, generator
  electrical dynamic type 2-7
  mechanical friction type 2-6,7
Buchholz relay 7-2,4
bus systems
  isolated phase 15-2,3,4
  non-segregated phase 15-2,4
  ratings 15-3,4
  segregated phase 15-2,3,4
bushings 7-1,2,4,6,7,8
bypass input transformer 14-3,6,8,9
cable shield connections 12-15,19
calibration, control system 11-2,4,6
carbon dioxide (CO_2) fire protection systems 2-8,9; 3-11,12
cast coil transformers 9-2,3,4,5,6,7
chassis grounds 12-16
chopped wave test 8-2
circuit breakers, UPS 14-9; 16-1,2,3,4,5,6,7,8,9
commissioning 17-1,2,4,5,6,7,8,9
  control system 11-1,2,4,5,6,7,8,10,11
communication protocol 10-8,9,10
computerized maintenance management systems (CMMS) 10-5
condition-based inspection outages 4-2
conservator system 7-8,9
constant-current battery charging 13-7,11
constant-potential battery charging 13-7,8
continuous off-line UPS 14-4,5
control system 10-1,2,5,7,8,10; 11-1,4,5,6,7,8,9,10,11
  lightning protection 12-5
  tests, at commissioning 17-4,7
coolers 2-8
cooling system, transformer 7-4
coordinating circuit breakers 16-2,3,6,8
corona discharges 6-11
crest factor 14-2
current transformer tests 3-4,5,9
current-limiting fuses 16-4,5
dielectric tests
  absorption test 6-11
  armature winding 3-4,5,12
  field winding 3-4,5,8
  transformer 6-1,2,3,12,13,14
dry transformers 9-1,2,3,4,5,6,7
dissipation factor test 6-9,10
distributed computer control system (DCS) 10-1,2,5,8,10; 11-1,9,10
Doble test 6-10
documentation of tests 17-2,5
down conductor 12-11
drawings, unit controls 11-3,4,5
dummy toggle switchboard 11-3,11
efficiency, UPS system 14-9,10

efficiency tests 17-4
generator 3-4,10,12
transformer 6-13
ANSI standard tolerances 8-1,2,3,4
IEC standard tolerances 8-1,3,4
electrical systems, review at start-up 17-6,8,9
electrical tests, generator, inspections
major scheduled inspection outage 4-2,5,6,7,9,10
minor scheduled inspection outage 4-2,3,4,5
electrically operated circuit breakers 16-4
electrolytes 13-1,2,3,4,5,6,7,8,9
electromagnetic core imperfection detector test (EL-CID) 3-6,7
emergency power 13-1,7
emergency shutdown control system 10-2
engineering review, at start-up 17-7
epoxy resin 9-2,3,4,7
equalizing charge 13-2,3,8,9,10
excitation current test 2-11; 7-3; 9-5
excitation system controls
automatic voltage regulators 2-12
excitation systems
brushless 2-11
static 2-11,12
fabricators, switchgear 16-9
factory tests
generator 3-1,2,3,8,13
transformer 6-3,7,12,13,14
unit control switchboard 11-2,3,4,5,6,8, 9,10
UPS 14-9
fall-of-potential test 12-10
ferroresonant type UPS 14-3
field check-out 17-1,2,5,6
documentation 11-6,7,8,11
field tests
at commissioning 17-2,3,5,6,8
generator 3-1,2,3,8,10,11,12,13
transformer
ground test 6-14
oil analysis 6-14
tank positive pressure 6-15
unit control switchboard 11-1,2,4,5,6,9, 10,11
filters, cooling water 2-8
fire extinguishing systems, generators 2-8
carbon dioxide (CO_2) 2-9
self-extinguishing insulation 2-9
water deluge 2-9
fire protection system 3-11,12,13
fire safety/hazard 9-1,4
float charge 13-2,7,8,9
flooded liquid electrolyte batteries 13-1,2,3,4,5,6
forward-transfer off-line UPS 14-4
frame size, circuit breakers 16-5,6
frequency, lightning stroke
relationship to ground impedance 12-14,15,16,19,20
relationship to time duration 12-2
frequency of inspection outages 4-2,7,8
functional tests, control system 11-2,4,11
generator circuit breakers 5-4
generator components 2-4
generator step-up transformers 5-1,2,3; 7-1
generators
semi-umbrella type 2-2,3,6,7
standard type 2-2,3,4,7
umbrella type 2-1,2,3,6,7
graphical display 10-5,7
ground conductor 12-5,15
ground connections, transformer 7-2
ground fault settings, circuit breakers 16-7
ground mat 12-5,9
ground resistance 12-10,11,14
ground systems 12-14
grounding, transformer 6-14
harmonic distortion 14-2,9
heat run test 3-9
high-potential test 3-3
high-pressure oil lift system 2-8
hot collar test 7-7,8
hot spots, bushings 7-4,8
hydrogen gas, from batteries 13-1,4,5,6,7
hydrostatic tests 3-4,5,8,11,12
ignition devices 6-9
impedance, ground conductor 12-10,14,15,16,19,20
impulse generators 6-7,8
impulse tests
chopped wave 6-3,5,6,7,9; 8-2,3
front-of-wave 6-3,6; 8-2,3
full wave 6-3,5,6,9; 8-2,3
lightning 6-3,6,7
reduced full wave 6-3,5,9; 8-2,3
tolerances 8-2,3
impulse wave form 12-17
index tests, turbine 17-3,4,11
induced voltage test 6-3,4,5,13; 9-6
infrared scan 7-4,8
input/output (I/O) cards 11-3,8
lightning protection for 12-13,18

inspection outages
  scheduled 4-1
    condition-based 4-2
    time-based 4-2
  unscheduled 4-2
installation procedures, batteries 13-1,10
instantaneous settings, circuit breakers 16-5,6,7
insulation
  power factor test 6-3,9,10,14; 9-7
  resistance test 6-10,11,14; 7-3,4
    armature winding 3-12
    field winding 3-5
insulation, transformer 6-1,3,4,9,10,11,14; 7-3,4,6
International Organization for Standardization (ISO) 10-8,9
international standards, transformer 8-1,2,3,4
inverter 14-3,4,5,6,7,8,9
isokeraunic maps 12-2,3
K factor 14-2,8
large hydro plants
  control system at 10-5,7
  protective relay at 10-5,7
lead acid batteries 13-1,3,7
  antimony batteries 13-3
  calcium batteries 13-3,7,10
  plante batteries 13-3
  special alloy batteries 13-3
leader channel resistance 12-2,3
lightning
  attraction systems 12-9
  cold lightning 12-2
  dissipation systems 12-7,8,9
  hot lightning 12-2
  protection 12-1,5,7,9,14
lightning impulse tests (see impulse tests)
line charging ratings 2-10
liquid-immersed transformer 6-1
load loss test 9-5,6
load rejection tests 17-3,4,11
logic grounds 12-16
long-time settings, circuit breakers 16-5,6,7
losses, load and no-load 6-1,4,9,10,13,14; 8-1,2
low-voltage switchgear 16-1,2,6
magnetizing test 3-6
main power transformers 5-1; 6-1,2,3
  common arrangements 5-2
maintenance tests, transformer 7-1,3
major scheduled inspection outages 4-2,5,6,7,9,10
man-machine interface software 10-5
manual backup controls 10-3
manual bypass switch, UPS 14-8
manually operated circuit breakers 16-4
manufacturer representative 4-7,10
  at commissioning 17-5
material test certificates 3-4,5,8
mechanical tests
  generator
    air gap 3-8
    bearing clearances 3-8
    brake operation 3-8
    cooling water flow 3-8,11,12
    hydrostatic 3-4,5,8,11,12
    moment of inertia 3-8,12
    rotational 3-8,11,12
    runaway speed 3-8,12
  generator, inspections
    major scheduled inspection outage 4-5,9
    minor scheduled inspection outage 4-2,4
  transformer
    leak test 6-13
    pressure test 6-13
    weld test 6-13
megohmmeter ("megger") test 3-5; 6-10; 7-3,4
microprocessor-based protective relay 10-7
  communication capabilities 10-10
  event storage 10-10
  multiple protective functions 10-10
  programmable setpoints 10-10
  self-calibration 10-10
minor scheduled inspection outages 4-2,4
mode switching, battery chargers 13-9
modems 10-5
  surge protection 12-19
molded case circuit breakers 16-1,2
National Lightning Detection Network (NLDN) 12-3,5
neutral-current wave shape 6-8,9
nickel cadmium alkaline batteries 13-3
nickel iron alkaline batteries 13-3
nitrogen system 7-2
no-load tests 17-3
  saturation test 3-4,8,10
non-linear loads 14-2,8
off-line UPS 14-2,4,5,8
oil analysis, transformer 6-14

oil tests, transformer
ASTM tests 7-5
dielectric test 7-5
dissolved gas analysis 7-5,8
on-line UPS 14-2,5,6,7,8
open circuit saturation test 3-8
Open Systems Interconnections (OSI) 10-8
outages 4-1,2,3,4,5,6,7,8,9,10
overload protection 16-7
partial discharge test 6-3,11; 9-6
performance test, batteries 13-11,12
personal computers 10-5
phase rotation 3-4,7,11,12
pilot cell 13-2,7,10
plant auxiliaries 5-3,5
plant control switchboard 10-1
polarity 8-4
polarization index test 6-11
polytetrafluoroethylene, in bearing pads 2-8
potential transformer 15-1,4
power factor 2-10
power factor test 6-3,9,10,14; 7-6,8
power factor tip-up test 7-3,4
power line conditioner 14-5
power supply 13-1,2,3,7,8,9,11; 14-1,2,3
powerhouse technical review 17-8
pre-commissioning tests 17-1,2
predictive maintenance components 4-2,8
pressure, stator core 3-6
pre-start-up tests 17-1,2
primary metal body 12-3
processing speed 10-2,7
programmable logic controller (PLC)
10-1,2,3,4,5,8,9,10; 11-1,3,8,9,11;
12-11,13,16,18,19; 17-10
grounding 11-3,8
software coding 11-3,8
surge protection 11-3,8
protective relays 17-2,7,9,10
bench test 11-5,11
induction disk 10-10
microprocessor-based 10-7,10; 11-11
solid state 10-9,10,13
pulse width modulation, UPS type 14-3
radiator, transformer 6-13
radio influence test 6-3
ratings
circuit breakers 16-5,6,7,8
generator
armature voltage 2-10
MVA 2-8,10
surge suppressors 12-17
rectifier 14-2,3,5,6,7,8,9
redundant PLC system 10-2,3
relays 10-1,3,4,5,9,10,11,12,13
repair outages 4-2
resident engineer 17-5,6
resistance temperature detector 10-5
resistance tests
armature 3-4
field winding 3-4
response time, control system 10-2,7
retrofitting LV switchgear 16-8
reverse-transfer on-line UPS 14-5,6,7
ring test 3-6
root mean square (RMS) asymmetrical fault current 16-5,6
rotor, generator
damper windings 2-5,6
field windings 2-5,6,11
poles 2-1,4,5,6
rim 2-4,5,6
spider 2-4,5,6
rotor inspection 4-3,4,5,6,10
salient pole machines 2-1,5
scheduling, of start-up tests 17-6
sealed maintenance-free batteries
absorbed electrolyte 13-2
gelled electrolyte 13-2
secondary metal body 12-3
selective coordination, circuit breakers 16-2
service test, batteries 13-11
setpoints 10-10,12
shaft, generator 2-4,5,6,11
shield wires 12-8,9,19,21
short circuit saturation test 3-4,8
short-time settings, circuit breakers 16-5,6,7
side flash 12-3,5,6
single-point ground systems 12-14,15,19
sizing batteries 13-4
sizing battery chargers 13-9
sizing UPS systems 14-6
skin effect 12-16
small hydro plants
control system at 10-2
protective relay at 10-5,11
software 10-1,2,3,5,7,10,12
soil ohmic resistance 12-10
solid-state protective relays 10-9,10,13
specific gravity, electrolyte 13-2,6,7,8,10
specifications 5-5,6; 10-7
commissioning tests 17-1,4,5,6,7,8
unit control switchboard 11-1,6,7,8
UPS 14-1,3,6,7,8
standards, international 5-6

start-up and commissioning 11-10; 17-1,2,3,4,5,6,7,8,9,10,11
start-up tests 17-1,2,3,6
static transfer switch 14-2,3,4,5,8,9
station auxiliary transformers 5-3,4
station power 16-1,2,4
stator, generator 2-4
stator core loop test 3-4,6,7,12
stator inspection 4-3,4,5,6,8
stator windings 3-3,5,6
  generator 2-4,5
step load change 14-2
substation 16-3,9
substations, lightning protection 12-4,8,9,10,13,14
sudden pressure devices 7-2
surge arrestor 12-13,16,18
surge protection 12-1,3,5,11,13,14,16,17,18,19; 17-8,9,10
surge protectors
  gas tube 12-17
  hybrid 12-17
  metal-oxide varistors 12-17
  silicon avalanche 12-17
switchgear 16-1,2,3,4,5,6,8,9
switching surge impulse test 6-6,7,13; 8-4
sync-check relay 11-3,6,8
synchronizing 11-3,5,6,11
synchronous condenser ratings 2-10
synchronous impedance test 3-9
systems operational testing 17-2
tan delta test 7-3
tank, transformer 6-13,15; 7-1,2,6,7,8,9
tapchanger 6-4,8
  on-load 7-9
telephone influence factor test 3-10,11,12
telephone lines, surge protection 12-19
temperature considerations, batteries 13-2,5
temperature indicators 7-2
temperature rise, dry transformers 9-2,4,5,6
temperature rise test 3-4,9,11,12; 9-5
terminal-voltage wave shape 6-9
test certificates 3-2,3,4,8
test sequence, transformer 6-5,13; 8-2,3
test specifications, generator 3-2,3,13
test specifications, transformer 6-1,2,3,13,14; 8-1,2
testing
  batteries 13-11
  bus systems 15-4
  protective relays 10-12,13
  UPS systems 14-9
tests
  at commissioning 17-4,9
  dry transformers 9-2,3,5,6,7
  transformers 7-1,3,4,5,6,7,8,9
  unit control switchboard 11-1,2,3,4,5,6,7,8,9,10,11
thrust bearings 2-4,7,8
transformers 7-1,2,3,4,5,6,7,8,9; 8-1,2,4; 9-1,2,3,4,5,6,7
  cast coil epoxy resin 5-5
  dry type 5-5
  liquid-immersed 5-1,2,3,4,5,6
transient protection (UPS) 14-8
transmission structures, lightning protection 12-8
turbine power output 2-9
turns ratio test 7-3
ungrounded specimen test 7-7,8
uninterruptible power supply systems (UPS)
  rotary type 14-1,2,3,4,9
  static type 14-1,2,3,4,5,7,8,9
unit auxiliary transformers 5-3
unit control switchboard 10-1,4,5; 11-1,2,3,5,9,11
unwatered tests 17-1,2,5,6
upgrade outage 4-2
upgrading switchgear 16-8
utility communications architecture (UCA) 10-9
vacuum pressure impregnated (VPI) transformers 5-5; 9-2
valve-regulated sealed lead acid batteries 13-1,2
visual inspection
  generator
    major scheduled inspection outage 4-5,9
    minor scheduled inspection outage 4-2,3,4
visual inspection, transformer 7-1
vital loads 14-2
Wagner's Method, lightning protection 12-9
watered-up tests 17-1,2,3,6
water-to-air coolers 2-8
wave impulse tests 6-6
waveforms 6-7
  deviation factor test 3-10,11,12
waveshapes 8-2,3,4
windings, transformer 6-1,3,4,5,8,9,10,13; 7-1
wiring, transformer 7-3,4,7,8,9
witnessing tests 11-4
zone of protection 12-3,4,5,6,7
zone selective interlocking 16-2